FACILITY SITING IN THE ASIA-PACIFIC

Facility Siting in the Asia-Pacific

Perspectives on Knowledge Production and Application

Edited by

Tung Fung, S. Hayden Lesbirel, and Kin-che Lam

The Chinese University Press

Facility Siting in the Asia-Pacific: Perspectives on Knowledge Production and Application

Edited by Tung Fung, S. Hayden Lesbirel, and Kin-che Lam

ISBN 978-962-996-406-1

The Chinese University Press
The Chinese University of Hong Kong
Sha Tin, N.T., Hong Kong
Fax: +852 2603 6692
+852 2603 7355
E-mail: cup@cuhk.edu.hk
Web-site: www.chineseupress.com

Printed in Hong Kong

Dedication

In memory of our teacher, Professor Lo Chor-pang

—Kin-che and Tung

To Ysabel and Adrian and in memory of their loving grandmother, Audrey Lenore Lesbirel

—Hayden

List of Contributors

Daniel P. Aldrich, Purdue University, USA

Jamie Baxter, University of Western Ontario, Canada

Chang-tay Chiou, National Taipei University, Taiwan

Tung Fung, The Chinese University of Hong Kong, Hong Kong

Te-hsiu Huang, Chung-Hua Institution for Economic Research, Taiwan

Kaoru Ishizaka, Okayama University, Japan

Howard Kunreuther, University of Pennsylvania, USA

Kin-che Lam, The Chinese University of Hong Kong, Hong Kong

Wai-ying Lee, The Chinese University of Hong Kong

S. Hayden Lesbirel, James Cook University, Australia

Virginia Maclaren, University of Toronto, Canada

Yasuhiro Matsui, Okayama University

Bruce Mitchell, University of Waterloo, Canada

Nguyen Quang Tuan, National Institute for Science and Technology Policy and Strategy Studies, Hanoi, Vietnam

Euston Quah, Nanyang Technological University, Singapore

Daigee Shaw, Chung-Hua Institution of Economic Research, Taiwan

Masaru Tanaka, Okayama University

Raymond Toh Yude, Ministry of Transport, Government of Singapore

Lai-yan Woo, The Chinese University of Hong Kong

Contents

1 Introduction 1
Tung Fung and Kin-che Lam

2 Facility Siting: The Theory-Practice Nexus 7
S. Hayden Lesbirel

3 Procedures for Dealing with Transboundary Risks in Siting Noxious Facilities 33
Howard Kunreuther

4 LULUs, NIMBYs, and Environmental Justice 57
Bruce Mitchell

5 Are Casinos NIMBYs? 85
Euston Quah and Raymond Toh Yude

6 Power to the People! Civil Society and Divisive Facilities 115
Daniel P. Aldrich

7 Site Selection of LULU Facilities: The Experience of Taiwan. 141
Chang-tay Chiou

8 Challenges of Managing NIMBYism in Hong Kong 169
Kin-che Lam, Wai-ying Lee, Tung Fung, and Lai-yan Woo

9 Community-Driven Regulation, Social Cohesion, and Landfill Opposition in Vietnam 183
Nguyen Quang Tuan and Virginia Maclaren

10 Reassessing the Voluntary Facility-Siting Process for a Hazardous Waste Facility in Alberta, Canada 15 Years Later 215
Jamie Baxter

11 Structural Model of Risk Perception on Landfill Site for Municipal Solid Waste. 231
Kaoru Ishizaka, Yasuhiro Matsui, and Masaru Tanaka

12 Compensation in Siting Hazardous Facilities: A Radioactive Waste Repository in Taiwan. 243
Daigee Shaw and Te-hsiu Huang

13 NIMBY: Environmental Civic Society and Social Fairness in China 257
Yang Yan

14 Conclusion 273
S. Hayden Lesbirel

1

Introduction

Tung Fung and Kin-che Lam

All societies require a full array of facilities to provide services and support for societal development. While some of these facilities may be greeted warmly by local communities, others are less welcome and are increasingly being rejected by those communities. This phenomenon is often referred to as Locally Unwanted Land Uses (LULUs) and the Not In My BackYard (NIMBY) dilemma. Facilities such as power plants, hospitals, highways, prisons, waste treatment facilities, landfills, incinerators, chemical waste disposal, and treatment plants are in demand in the Asia-Pacific region. The need for these facilities is of little dispute among citizens, particularly at the national level. While planners and decision makers need to determine where to locate these facilities, it has become an increasingly daunting task.

While such projects can bring significant gains to both local and national communities, the negative spillover effects, including environmental, social, economic, and health impacts imposed on the local communities are often insurmountable. More often than not, local communities raise serious concerns that often lead to protest and opposition. This can result in project delays, increased developments costs, and even cancellation of projects. Certainly, local communities are vulnerable to the risks associated with facility siting and the challenge for decision makers is to find ways to provide sound communication, effective assessment, and management of the risks involved.

The mismatch between who gains and who loses from the development of projects leads to conflict and, hence, the siting of projects requires a conflict resolution process. The success of this process often rests on the ability of promoters to build up trust and equity in a situation where there is considerable tension amongst the interest groups. Different countries and governments have attempted a variety of methods for facilities siting, with some adopting a more "decide-announce-defend" approach, while others attempt a more voluntary siting approach. Independent of the approach, there have been varying degrees of success, and cases of failure, in particular, draw intense media attention and can exacerbate the problem. Evidence from the Asia-Pacific suggests that the siting problem has emerged independent of the form of government or level of economic development of nations in the region. The inability to manage conflicts in a timely fashion has serious implications for the achievement of national and regional policy priorities.

Research on facilities siting has both academic merit and practical relevance to various stakeholders. Since the 1980s, there has been a significant growth in the literature that deals with the origins and management of conflicts involved in siting facilities that are perceived to be public "nasties." This book adds to the literature in three ways. First, it evaluates the extent to which a focus on siting in the Asia-Pacific can enhance our knowledge of siting theoretically and comparatively. Much of the facility siting literature originates from experience in North America and Europe. Many of the books on siting continue to focus on Western experience, although there have been some works on Asian experience (mainly in Japan and Taiwan). This book, by explicitly focusing on Asia-Pacific experience and covering countries such as China, Singapore, and Vietnam that have not been covered adequately in the literature, seeks to make a major contribution to the growing comparative siting literature. Second, it explores the extent to which the literature provides insights to policy practitioners involved in managing siting disputes. The siting literature is highly policy-relevant. Criticisms of bringing *policy relevance* back into social science do not hold up in the case of siting. Yet, ironically, there is little analysis on how the literature can assist policy makers in developing better siting policies and effectively managing siting conflicts. Third, it explores the scope of the subject matter covered by the siting literature. Much of the literature makes two critical assumptions. The first is that it presumes the only conflicts that matter are those involved between *host communities* and developers, whether they be private or public or some combination. The second is that it assumes the siting issue ends during the *preconstruction* stage. This book challenges both assumptions and stresses the importance of neighbouring communities in siting conflicts and the need to consider *postconstruction* conflicts, both of which can have significant implications for siting processes and outcomes.

S. Hayden Lesbirel, in the next chapter, provides the first extensive survey of the siting literature since the mid-1970s and focuses on the relationship between the production and use of knowledge in facilities siting. It suggests that the siting literature has developed into a fully fledged literature that uses a full range of theoretical and methodological approaches to explore siting conflicts, and has produced a variety of middle-range theories to explain the origins and management of those conflicts. The literature is highly policy-relevant and can provide not only important conceptual insights to siting practitioners in terms of basic perspectives and orientations, but can also offer instrumental insights in strategic and functional terms. The challenge for the literature in the future will build on these achievements and address several theoretical and empirical shortcomings in ways that seek to fulfil the needs of siting practitioners.

In chapter 3, **Kunreuther** examines ways to better manage the trans-boundary risks associated with LULU. By understanding the nature of the problem from different stakeholders, this chapter suggests a framework for evaluating alternative siting strategies. It also explores how a siting authority could achieve consensus-building and examines the role of mitigation measures and compensation in the process. It then suggests a set of issues that need to be addressed regarding the involvement of the interaction among policy makers, risk management institutions, and the public in dealing with transboundary risk problems facing the public and private sectors.

Mitchell in chapter 4 focuses his discussion on the relationship between unwanted facilities and the concept of environmental justice. The ways in which governments in North America have interpreted and used environ-mental justice as one means to address issues related to LULUs and NIMBYs are examined. Furthermore, by examining examples in Canada, approaches (both traditional and voluntary) used in the siting process of LULUs are also explored. The chapter specifically identifies the importance of transparent principles, engagement of local communities from the outset in the decision process, innovative procedures (reverse Dutch auction), and opportunities to overcome mistrust.

LULUs certainly include a variety of facilities. Although a casino may not be a typical NIMBY example, the introduction of a casino into any society is controversial, as one would easily debate on the potential economic gain and job opportunity created versus the negative externalities and social costs. **Quah** and **Toh** in chapter 5 discuss this issue based on two casinos in Singa-pore and provide a forceful argument on the importance of public goods provision.

With few investigations on broader patterns by which authorities locate LULU facilities, **Aldrich** in chapter 6 calls for a reorientation of scholarship on land-use conflict to better capture methodological advances in the social sciences, including large-scale data analysis and political geography to bridge the knowledge gap. The new analytical tools uncover the strength of local networks and social capital in the siting process. The paper concludes that civil society has an important role to play in determining the success of implementing new technologies, and thus highlights the importance of local communities' characteristics in determining which policy tools states use and their likely effectiveness in siting.

Following on from these conceptual developments in siting analyses, **Chiou** in chapter 7 examines which factors contribute to effective siting of large-scale projects in Taiwan. The study reveals that siting syndrome emerges in the field of Taiwan's site selection and construction of electric power stations and solid waste incinerators. Chiou calls for greater sensitivity towards the influence of *noneconomic* factors on siting process rather than placing too much attention on the effect of *economic* factors, which he believes has traditionally been the case in the literature. Noneconomic factors in the context of Taiwan are associated closely with the lack of public participation, credibility deficiency, and local politics. The chapter suggests that planners should use the community-governance approach to resolve the dilemma of siting facilities, with more emphasis on noneconomic factors.

Moving from Taiwan to Hong Kong, **Lam et al.** in chapter 8 attempt to determine how the public views LULU facilities and whether its perception of risks is related to the type of facility. The chapter elucidates how NIMBYism has arisen in the specific political, social, economic, and geographical context of Hong Kong and explores how siting conflicts might be resolved. The study also indicates that despite the concentration of LULUs in Tuen Mun, a district with a disproportionate share of these projects, local residents are not keenly aware of these facilities. This paper also argues that monetary compensation is of limited effectiveness in reducing public resistance.

With inadequate enforcement of environmental regulations by the authorities, community-driven regulation (CDR) or informal regulation is an alternative measure used to resist public facilities, such as landfills, after they are sited. In chapter 9, **Nguyen** and **Maclaren** examine four landfills in Vietnam that experienced significant opposition from local communities and assessed the measures taken by the local communities and the effectiveness of CDR. The importance of social cohesion, social capital, and their relationship is also discussed. This chapter indicates that a more formal mechanism is needed to involve the public in siting and operation of noxious facilities to avoid community opposition in Vietnam.

By focusing on benefits rather than risks that could be brought by hazardous facilities, voluntary siting is preferred over traditional, more coercive, methods. **Baxter** in chapter 10 evaluates factors leading to the success of voluntary siting. The chapter explores the case of Swan Hills, Alberta, Canada, which is often portrayed as one of the earliest and most successful cases of voluntary siting in North America. However, he points out that siting cannot claim to have achieved justice as Swan Hills was not in a disadvantaged or vulnerable bargaining position. In fact, the perceived fairness of the original siting process is the strongest predictor of facility-related concern, both in the host and neighbouring communities. Implications for the viability of voluntary siting, the appropriate role for informed consent, and the associated role of scale are also discussed.

Ishizaka et al. in chapter 11 analyze factors relevant to the acceptance and risk perception of landfill site for municipal soil waste in Okayama city, Kurashiki city, and Yoshinaga city in Japan's Okayama prefecture. The study reveals that risk perceptions and trust in technology and standards are factors that influence the acceptance of landfills; while trust in technology and standards, and trust in response to accidents are the factors affecting the risk perception. The study reconfirms the importance of an open-door policy and daily communication between citizens and local government in the siting process.

In chapter 12, **Shaw** and **Huang** investigate siting a low-level radioactive waste repository in Wu-chiu, and find the way in which the compensation is provided. Fairness, trust in the developers, siting procedures, and income are important factors characterizing the public's perception of and attitudes toward the facility, while trusting in negotiators is the key way residents can make their decisions in Taiwan. The result is compared with cases in the United States, Switzerland, and Japan. The chapter also identifies differences among the cases in relation to social capital, including civic duty, social pressure, and trust in the developers. It concludes that social capital is an important aspect of public opinion in relation to the NIMBY phenomenon.

Yang, in chapter 13, studies the emergence of environmental nongovernmental groups in siting NIMBYs in China. Based on two cases, this chapter concludes that, given the rapid economic development and adoption of more open and transparent policy, environmental nongovernmental groups are able to exert their influence by stressing environmental rights and social justice, leading to a postponement of a key hydroelectric development in Nujiang and the resiting of a chemical plant in Xiamen.

By exploring siting problems in the context of the relationship between the knowledge production and use in the Asia-Pacific, this book suggests expanding the literature's subject matter to more fully incorporate the impact of neighbouring communities and post-siting conflicts on understanding the origins and management of siting outcomes. Doing this will enhance the conceptual and instrumental utility that the literature offers to stakeholders involved in siting processes not only in the Asia-Pacific but elsewhere as well.

2

Facility Siting: The Theory-Practice Nexus

S. Hayden Lesbirel

INTRODUCTION

Being a siting practitioner is not an easy task.

The siting and development of a range of projects, such as waste repositories, prisons, energy facilities, airports, and industrial projects have and continue to be a lightning rod for social and political conflict in all nations. States and firms may need to develop such projects to provide a range of social and economic benefits for the national community. Yet, while local community interests may agree with the broader social need for these projects, they often oppose them vigorously as they perceive them as imposing significant costs, such as environmental degradation, unacceptable levels of risk, and disruptions to social relationships, on their communities. These responses often create considerable conflict and can delay or even cause abandonment of facility plans.

Siting is clearly a case of contentious politics that can impose significant costs on stakeholders. Siting is a significant policy issue that impacts the achievement of state, corporate, and community objectives. Disputes have been costly for states, particularly where projects are needed for national technological, economic, and security objectives. Conflicts have often been costly for developers and include increased uncertainty over capital cost escalations due to inflation and interest repayment burdens. They have also been costly for communities as they have, for example, altered existing social and political relationships and levels of social capital within those communities. Such outcomes might be beneficial, since the grounds for opposition might be well based, but they may be, in other cases, socially undesirable since failure to site such facilities might carry opportunity costs that are felt by other communities.

Practitioners confront a complex range of information, much of which is incomplete and ambiguous when contemplating facilities siting. They may have incomplete information about the magnitude of changing societal needs for projects they are planning. They may not be clear about the character of the stakeholders with whom they will need to negotiate in order to win agreement for the project siting. They are more than likely to be uncertain about the preferences and underlying motivations of these stakeholders and how they will respond to siting processes. They may have conflicting information about which strategies and policy tools might work and which ones might not. Furthermore, in many cases, as the stakes involved in these conflicts are large, such as with capital expenditures, getting any of these things wrong can be extremely costly for stakeholders.

Reflecting the increasing importance attached to siting as a social and policy problem, there has been a growing literature on the *production of knowledge* about the origins and management of conflict in the development of unwanted projects. Over the last thirty years or so, social scientists, including economists, geographers, historians, political scientists, sociologists, and psychologists, have developed a range of theoretical explanations that seek to account for recurring patterns of behaviour and discourse in siting controversies over time and across space. They have offered diverse explanations of siting processes, issues, and outcomes using a wide range of political, economic, demographic, and technological variables. They have used various competing theoretical and methodological approaches in seeking to explain and interpret siting controversies. While the field is relatively new, it has developed into a rich multidisciplinary field of inquiry.

Contemporaneously, there has been a growing literature on the *utilisation of knowledge* by a range of social scientists in a variety of policy fields, both domestic and international. This literature has sought to explain the relationship between knowledge production and utilisation in policy processes. It has sought to develop a range of models to investigate the use and impact of research on practical politics and policy and the processes through which knowledge production finds its way into knowledge utilisation (Stone et al., 2001). One recurring theme in this literature relates to understanding and analysing the factors, such as the cultural gap between scholars and practitioners, the validity and reliability of research, and the ways in which research is useful for practitioners involved in practical political processes.

How is it possible to make sense of this diverse scholarly siting literature and its possible usefulness to siting practitioners? What have been the major developments in the siting literature and what does the evolution of literature tell us cumulatively about siting? Have siting scholars developed theoretical explanations that are robust? To what extent have scholars of siting left their ivory towers and produced explanations of siting processes and outcomes that are of practical relevance to stakeholders? If so, what does the knowledge utilisation literature say about the potential utility of scholarly literature for practitioners? How can we understand the ways in which the production of knowledge can be used by siting practitioners? This paper brings together these two literatures and investigates the relationship between siting theory and practice.

KNOWLEDGE PRODUCTION

The social science literature on siting represents an effort by social scientists in knowledge production by seeking to account for issues, processes, and outcomes involved in the siting of a wide range of facilities. Social scientists are not unified in positions and approaches to the production of knowledge. They have differing ontological views (whether there is a real world out there or whether that world is socially constructed), epistemological orientations (how do we know what we know about the real world), theoretical views (what are appropriate theories and what variables should be included), and methodological views (what are the most useful methods for understanding the social world). These differences produce varying, yet important, explanations of siting controversies and their management.

Subject Matter

The facility siting literature covers a wide range of projects, both domestic and international. Of particular importance has been coverage of controversial, high-risk projects such as nuclear power plants, waste repositories, and large-scale industrial projects such as chemical facilities. It also has analysed a host of other projects that might appear at first sight to be less controversial, but that can actually generate significant community conflict. These include libraries, wind farms, hospitals and other medical facilities, museums, movie production facilities, and bridges. While domestic projects constitute the bulk of the literature, it has also covered projects that cross borders of sovereign states or whose impacts are perceived to cross those borders. These include nuclear facilities that are located close to other states (Loefstedt, 1996) and oil and gas pipelines that transit two or more states (Hansen, 2003).

Developing projects generally requires a planning stage (site selection) and an implementation stage (public acceptance, licensing, construction, operation, decommissioning). While it is often not possible to separate these stages analytically because of overlap, it is possible to highlight the key features of them. Site selection typically involves the use of technical criteria, such as the existence of suitable terrain, and the availability of resources such as land, to establish a pool of least-cost candidate sites from which a site would be eventually selected. Public acceptance generally involves political processes aimed at securing community acceptance of projects at selected sites. Licensing involves a regulatory process of government seeking to balance the expected social and economic benefits and risks of facilities. The subsequent stages of project implementation typically involve optimisation processes, whereby developers seek to minimise construction, operation, and decommissioning costs (Lesbirel, 1998).

The earliest comprehensive review of the siting literature focused on planning and site selection, particularly the use of numerical methods to select least-cost sites on which to develop projects (Jopling, 1974). It surveyed the site selection literature focusing on the use of technical criteria in site selection such as the existence of flat and stable terrain, the availability of cooling water, a relatively low population density (particularly for nuclear plants), accessibility to transportation routes, and proximity to major load centres. It also investigated the use of numerical methods (attaching numerical scores to site selection criteria) in establishing pools of least-cost candidate sites from which a site would be eventually selected.

Since that time, the siting literature has grown significantly and is still growing. Several books have been written about siting and articles on siting can be found in many journals. The bulk of this growth in the literature has focussed on the social and political aspects of siting, particularly during the public acceptance stage. This has reflected the increased political difficulties of siting and an attempt by social scientists to explore the origins and management of siting conflicts. Examinations of the other stages of implementation have tended to take a back seat in the literature. This is perhaps not surprising, given that the regulatory and economic optimisation processes involved in implementation (including decommissioning) only become relevant once public acceptance has been achieved, in whatever form. It is timely to review the literature in the context of social science.

Theoretical Approaches

We can understand this growing literature by reference to the ontological and epistemological approaches to the production of knowledge (Hay, 2002). Ontology is a theory of being. The key question is whether there is a real siting world out there that is independent of our knowledge of it. For instance, are there *fundamental* differences between the risks of nuclear projects and wind farms? Foundationalists might agree and argue that these differences persist across space and time and that these differences provide a critical foundation on which to explore siting disputes. In contrast, antifoundationalists would most likely argue that these differences are not *fundamental*, but are particular to different times, cultures, and circumstances. They would argue that there are not objective differences between the risks of different projects but that any differences are socially constructed and that these social constructions have a significant bearing on siting processes.

The ontological positions that social scientists adopt influence their epistemological positions on scientific claims or what we can know about the siting world and how we can know it. For example, are there real or objective relationships between risks and the degree of difficulty in siting different projects, and can we observe those relationships directly? There are three general positions. Positivists would argue that it is possible to understand this relationship through theory and to test that relationship by direct observation (Halfpenny, 1982). Interpretivists would argue that the world is discursively or socially constructed (Foucault, 1977), that it is not possible to observe siting phenomena directly, and that it is crucial to identify the subjective interpretations or meanings attached to the risk-siting difficulty relationship. Realists sit between positivists and interpretivist (Sayer, 1992). They would argue that there is a real and objective risk-siting difficulty relationship, but that there are also deep but unobservable social, economic, and ideological structures that would account for differences in siting phenomena.

These competing ontological and epistemological positions underpin a diverse range of theoretical and methodological approaches used in analysis of siting conflicts. I use a framework contained in Marsh & Stoker (2002) to review briefly the siting literature in terms of these competing positions and approaches in the social sciences. It treats institutional, behavioural, and rational choice theory as the dominant foundational approaches contained in the siting literature. It categorizes feminist theory as a foundationalist approach, but recognizes the increasingly strong tendencies towards antifoundational approaches. Interpretive theory is classified as the antifoundational approach. Finally, it considers normative theory from the perspective of both ontological approaches.

Institutional approaches to siting are concerned with exploring the institutions, rules, procedures of the political system, and the impact on siting politics through organized knowledge that is theoretically informed. They cover organisational structures and relationships between different arms of government, as well as the impact of policy networks on politics and policy outcomes. An important perspective is that they treat government not in organisational terms, but as an association of heterogeneous political actors in their own right with their own political interests. Weingart (2001) shows that the nature and structure of state institutions influenced the inability to site low-level radioactive waste sites in New Jersey in the second half of the 1990s. While the state government was creative and flexible, and while there was some organized resistance, there were significant bureaucratic constraints resulting from different policy goals and overlapping jurisdictions that ultimately prevented the state from managing siting conflicts. McAvoy (1994) highlights how the siting of waste facilities in Minnesota was problematic even when strong elite policy networks comprising government, industry, and environmental groups (led by the Sierra Club) agreed that the solutions were acceptable. While local citizens' groups did not impact on state autonomy, they were able to derail siting processes and significantly influence the capacity of the state to achieve its siting objectives.

Behavioural approaches seek to explain political behaviour of stakeholders involved in siting conflicts through the development of falsifiable statements that are then tested against the evidence. They emphasise the question of why people at the individual and aggregate levels behave politically the way they do and how we account for their behaviour in siting conflicts. They focus on observable behaviour and use of theory and explanation to develop a causal account of the relationships between behaviour and siting, using systematically all the relevant evidence. Lober (1995) discovers that behavioural opposition declines more quickly with distance than attitudinal resistance, suggesting that self-interest rather than attitudes are crucial in explaining varying behavioural responses to siting facilities. Dear (1992) and Hunter and Leyden (1995) provide some evidence to show that more educated, younger residents with higher incomes are more likely to oppose facilities. This provides some explanation as to why some observers have suggested that project developers seek to locate unwanted projects in poorer areas that are characterized by less educated and nonwhite residents (Bullard, 1990; Been, 1994; Kruize et al., 2007).

Rational choice theory focuses on political choices made by rational, self-interested individuals and seeks to develop general laws regarding these choices. It argues that political actions by stakeholders involved in siting can be understood in these terms. It assumes a rational capacity by stakeholders to choose among alternative courses of action the one that they believe is likely to have best overall outcome. It therefore seeks to explain political choices that stakeholders make and the resultant outcomes in terms of courses of action or strategies given preferences over goals and beliefs about what influences the preferences of other actors. Frey and Oberholzer-Gee (1997) use rational choice theory to explain the crowding-out effect of monetary compensation. They found in a Swiss case that the level of acceptance of facilities drops when compensation is offered because intrinsic motivation is partially destroyed (reducing the option of indulging in altruistic behaviour) when price motivations are introduced. Hamilton (1993, 2005) uses a discrimination model to explore locational features of projects. He argues that profit-maximizing firms are likely to select sites where there are low income and education levels, as there is a relatively low willingness to pay for the environment. As firms care about political opposition, they prefer to locate in minority areas, as a lack of weak collective action requires relatively less internalization of negative spillover effects.

Feminist theory stresses the impact of and challenge to the structure of patriarchy (rule by males) as a form of power in political processes. It argues that there has been a gender blindness in foundationalist approaches and explores the nature and importance of gender in understanding siting conflicts. The theory expands the political debate from the *public* to the *private* sphere. Given evidence that women tend to attribute higher risks to siting-related activities than males (Slovic, 2000), it provides insights into the relationship between gender, power, and the management of siting conflicts. Brown and Ferguson (1995) argue that women constitute the majority of both the leadership and the membership of local toxic-waste activist organizations, and show how women activists transcend private pain, fear, and disempowerment and become powerful forces for change by organizing against toxic waste. Bantjes and Trussler (1999) agree, arguing that the fit between the community health focus and women's traditional role (the motherhood effect) enables women to play a central organizational role in antiwaste movements. They conclude that women have greater structural availability than men do in fighting toxic waste projects as they are less likely to be in the labour force and that housewife activists can form stronger local female networks, based on ties of kinship and domestic labour, to provide powerful opposition.

Interpretivist theory represents the antifoundationalist approach, and generally rejects foundationalist approaches to social science. Dismissing the notion that a real world exists, it focuses on the structuring of social meaning as central siting controversies and argues that the system of meaning (discourses) shapes the way people understand political activity involved in siting. It sees political actors, institutions, and practices as only making sense within a particular discourse in terms of the use of language, symbols, and the structuring of siting debates. Siting conflicts are viewed in terms of the production, functioning, and changing of discourses relating to key aspects of siting processes, such as equity and identity. These conflicts are conflicts among different forces trying to impose ideas (structures of meaning) on each other. Hubbard (2005) applies discourse theory to the siting of an asylum facility in England. He clearly demonstrates how opposition could create a dominant discourse between self (local citizens) and others (asylum seekers) by emphasizing a social construction of the "other" as a burden on the community and a potential security risk, thereby threatening the identity of the English countryside. Haggett and Toke (2006) show how opponents to wind farms in Wales were able to dispel claims of NIMBYism by developing a discourse that challenged wind farms as clean technology by appeals to the notion of intrusion into unspoiled areas and the use of the language such as wind energy power station (image of large factories with smoke) as opposed to wind farm (images of being part of the countryside).

Normative theory concerns the discovery and application of moral notions to siting practice. It explores the goals, values, and processes of society that should be pursued such as equity (equal distribution of benefits and burdens), liberty (rights of government to interfere with choices by local communities), and efficiency (maximising siting outcomes with minimum costs). In short, it addresses a central question of what ought to or should be and examines alternatives open to society by elaborating a "best blueprint for society." Foundationalist and antifoundationalist approaches all have a range of normative perspectives. Institutional, behavioural, and rational choice approaches generally stress the need for approaches to siting that are fair, workable, just, transparent, and legitimate, as well as the importance of institutions in achieving these normative goals. The feminist literature stresses gender and other inequalities in terms of the siting of projects. Interpretivists stress the discursive aspects of siting and view siting conflicts as importantly being a contest of competing ideas and discourses, all of which have normative foundations.

It is important to note that these competing theoretical perspectives are not mutually exclusive. The literature contains a variety of examples of multi-theoretic approaches to examining siting conflicts that attempt to provide more theoretically integrated approaches to understanding this issue. Hecht (1998) uses both institutional and behavioural analysis to explore the interactions between fragmented institutional decision-making and the behaviour of stakeholders in managing siting conflicts in rural France. Sakai (2005) combines social choice and normative theory to develop a formal model of siting that posits site selection in a way that maximizes social welfare to share the value equally through monetary compensation, and that such an approach would be robust to strategic manipulations. Haggett and Toke (2006) explore wind-farm siting conflicts in England and Wales and show how siting discourses related to the behaviour of key stakeholders.

Methodological Approaches

The literature employs competing methodological approaches to the analysis of siting conflicts and their management. The literature is replete with the use of qualitative analyses that have generally used case studies and narratives (often based on interviews and focus groups) to explore siting conflicts. These have been particularly useful in examining siting decision processes and outcomes; policy tools that states use in managing siting processes, motivations, strategies supporters and opponents employ in siting disputes, understanding the experiences of stakeholders in siting conflicts and the meanings they attach to these experiences, and in drawing attention to the broader social, political, and historical contexts in which siting conflicts occur.

The use of quantitative approaches is abundant in the siting literature. These methods include univariate, bivariate, and multivariate techniques to provide statistical insights into siting conflicts. They seek to explore statistically relationships among dependent and explanatory variables and assess the strength of those relationships, using both experimental and observational data. They have been useful for analysing the socioeconomic locational characteristics of projects in terms of environmental justice (Been & Gupta, 1997) and levels of social capital (Aldrich, 2007), the relationship between compensation offers and the crowding out of civic duty in siting facilities (Frey & Oberholzer-Gee, 1997), the relationship between auctioning strategies and compensation costs in siting (Quah & Tan, 2002), and the relationship between public acceptance times and the structure of the bargaining environment (Lesbirel, 1998).

Finally, the siting literature has also used comparative methods to further enhance our understanding of siting processes and outcomes. These methods aim to explore explicitly and systematically differences and similarities between siting processes and outcomes, and can involve intra- or intercountry comparisons, both across space and over time. Much of the earlier siting literature focussed on North American cases, although over time the country coverage has expanded to include a wider variety of settings in Europe, the Asia-Pacific, and elsewhere. Many of these studies are comparative in the sense that they compare different siting outcomes within these nations. An important development has been the increase of intercountry comparative analyses, including alternative siting strategies in the United States, Canada, and other advanced nations (Rabe, 1994; Munton, 1996), failure and success in siting

(Vári et al., 1994), transaction costs and institutions in industrialised nations (Lesbirel & Shaw, 2005), and state management of civil society in advanced nations (Aldrich, 2007).

Explanatory Utility

The range of theories used in the siting literature represents attempts by social scientists to account for recurring processes, issues, and outcomes involved in facility siting. The extent to which these theories are useful in explaining the real world of siting will be importantly determined by the empirical validity of causal relationships among the variables contained in those theories, and the extent to which the theories explain siting phenomena in general terms.

The literature has identified a wide range of variables to understand the origins and management of siting conflicts. These include: risk, trust, distribution of burdens, demand for environmental quality, compensation and mitigation, legitimacy, public participation, power, political party structure, social capital, strategy, and the like, and the literature has explored the empirical relationships between these variables and siting difficulties. For instance, Jenkins-Smith and Kunreuther (2005) explore the relationship between the use of compensation and changing degrees of opposition for projects of differing perceived risk levels in the United States. They find there is likely to be less resistance to the use of compensation for projects that are perceived to be less risky. Kraft (2000) stresses the importance of policy design in siting nuclear waste repositories in the United States and Canada. He concludes that Canada had adopted a more deliberate pace of policy development (including extensive public participation), while the United States had adopted a more rushed pace with respect to its policy development.

While unicausal explanations in the social sciences might provide useful partial understandings of siting processes and outcomes, the literature has developed sophisticated multicausal models that highlight interactions among independent variables. Such models generate better explanations of siting conflicts. Rabe et al. (2000) explore the relationship between voluntary siting strategies and trust, legitimacy, and risk, and show how siting became derailed when authority shifted from a public to a private implementing agency. Aldrich (2007) explores the relationship among demographic, political, and civil society variables and probable siting outcomes in Japan, demonstrating that state-planned projects are more likely to be located and implemented in communities where civil society is less concentrated and relatively weak.

An important feature of the siting literature is that there is no general theory of siting, but rather a variety of middle-range theories. General theories are wide in the scope of their general applicability and are characterized by little conditionality in their conclusions. They explain broader patterns of behaviour and discourses that persist, both across space and over time, with the use of a relatively small number of explanatory variables. The siting literature has not yet produced general theories such as positivist or interpretivist theories of siting. In contrast, the siting literature is replete with a variety of middle-range theories that are narrower in scope and have more conditionality in terms of their conclusions. Such theories tend to be more problem-oriented and focus on a specific set of issues, strategies, policy instruments, and the like in specific social and historical contexts. While they are related to more general social science theories and provide important insights, they are often based on a limited number of observations or cases and tend to produce contingent generalizations (Lane, 1990; Wilson, 2000; Jentleson, 2002).

Cumulatively, the siting literature has explored the origins and management of conflict in the siting of a wide range of facilities. It covers a full range of theoretical and methodological perspectives, although the literature is dominated by foundationalist approaches. It identifies a key set of variables and, importantly, demonstrates reasonably well how the interrelationships among these variables influence siting processes and their outcomes in complex ways. These analyses have enhanced our theoretical knowledge of different aspects of siting in a host of social, political, and historical contexts. While the literature has not developed general theories of siting, it has developed a range of middle-range theories that are important in understanding issues, strategies, and policy instruments.

KNOWLEDGE UTILISATION

A continuing theme in the knowledge utilisation literature relates to the nature and extent of a cultural gap between academia and policy practitioners (Stone et al., 2001). This is often referred to as a *two communities* model (Caplan et al., 1975). One view is that this gap is due to the different ways that the camps produce knowledge. Scholars generally see knowledge as deriving from theory. Policy practitioners generally view knowledge as stemming from experience and common sense based on their involvement in real-world social and political processes. One observer has gone so far as to say that academics are from Mars while policy makers are from Venus (Birnbaum, 2000). This model posits that that academic and policy communities are distinct, that there is little interaction between the two, and that there is limited use of scholarly knowledge by policy practitioners (Caplan et al., 1975; Booth, 1988; Eriksson & Sundelius, 2005).

Policy practitioners, like academics, have theories of their own relating to management of siting controversies. Practitioners have theories that guide them in identifying goals in siting processes and considering, evaluating, and choosing alternatives courses of action to develop projects. In doing this, they have to decide, amongst other things, which information to use, which stakeholders matter, which events need priority, which strategies to employ and in what order, and which policy instruments and in which combination they will use to achieve their goals, whatever they may be. These theories might derive from insights in the scholarly literature, previous experience in the siting of facilities by them or others, rules of thumb, or some combination of these.

Indeed, Hamilton (2005) notes that siting policy debates amongst practitioners in the United States can be understood in terms of competing theories to which policy practitioners subscribe. He observes that some decision-makers expressed a preference for an approach that generally left siting in the hands of private developers but specified a process of explicit negotiation between developers and communities, buttressed with compensation mechanisms that sought to offset negative expected social and environmental costs to local communities. He notes others advocated an approach that involved centralised decision-making power where the state had the authority to initiate siting processes and would have preemptive powers whereby it could override the zoning powers of local governments. Hamilton argues that these policy debates centered on whether states approached siting conflict management through a market model or a firm's decision-making model.

How can we evaluate the potential utility of scholarly research for siting practitioners? The knowledge utilisation literature provides a useful entry point. It identifies two major uses of research. The first is *conceptual* use of knowledge and the ways in which it can assist practitioners in the basic orientations and broader perspectives toward resolving social and political problems. The second is *instrumental* use of knowledge and concerns the ways in which the literature can assist practitioners in more strategic and functional ways (Caplan et al., 1975; Weiss, 1977; Jentleson, 1990; Walt, 2005). I apply these notions to evaluate the ways in which the siting literature can be helpful to siting practitioners.

Conceptual Utility

The history of the siting literature reflects a major paradigm shift during the 1980s and 1990s. As a result of the emergence of siting difficulties in the 1970s and 1980s in democratic nations, there was a basic change in awareness and a theoretical reorientation from a position that stressed coercive approaches to siting to one that emphasized more participatory, democratic approaches. The use of numerical least-cost approaches to site selection was closely associated with DAD (decide-announce-defend) approaches to managing siting conflicts. Developers, after selecting least-cost sites, would either seek to ride out community opposition or attempt to override community interests. Typically, secret discussions would occur between developers and political and other commercial elites in local communities with no public consultation. Developers would obtain relevant preliminary construction and other licensing permits. Once the siting proposal became public (either by accident or by intentional leaks), developers would seek to ride out any community opposition that emerged. Where this opposition was perceived to be strong or likely to become more intense, local governments would also seek to override that resistance through the use of zoning and other laws such as Eminent Domain (Kunreuther, 1995; Munton, 1996).

As the literature demonstrates, such coercive approaches to siting have generally not worked for some time (Kasperson, 2005), although Aldrich (2007) provides some qualification, suggesting that states do use coercive methods in siting some facilities. In democratic countries, communities are generally powerful enough to delay or stop the development of projects that they perceive to be risky. Many states still have the legal and constitutional authority to impose environmental burdens on community interests (through the use, for example, of Eminent Domain). Yet, increased demands for more voluntary and democratic processes, power sharing, and transparency, coupled with more awareness of environmental risks, equity issues, and mistrust in public institutions, have effectively meant that communities have veto power over project placement decisions. As Morell and Magorian (1982) conclude, governments can strip away the legal power of communities, but they cannot strip away their political power.

An important feature of the literature is that it is highly policy-relevant and has provided overarching perspectives that seek to assist practitioners in siting conflict management. The most seminal in this regard is the Facility Siting Credo (Kunreuther et al., 1993). The fundamental theoretical orientation of the Credo is that the key features of siting conflicts are disagreement over values and goals, a tendency to wish to maintain the status quo, and a lack of trust. Based on this, the Credo suggests a range of guidelines with the aim of achieving a more deliberative, workable, and fair siting process for all stakeholders. The Credo has formed the basis for subsequent policy-relevant siting research. Many analyses of siting have tested the validity of the Credo or have used it as a basis, either implicitly or explicitly, for developing further analysis and practical recommendations.

Particularly noteworthy is the development of a stepwise approach for nuclear waste-facility siting that draws heavily on the Credo (OECD, 2004; Pescatore & Vári, 2006). This approach stresses the reversibility of decisions after reconsideration of one or a series of steps at various stages in the siting process. The key theoretical perspective relates to participatory democracy and new forms of risk governance and, in particular, asserts that decision-making should be open and provide the flexibility to adapt to contextual change, that social learning should be facilitated, and that there should be public involvement in siting processes. Importantly, the study argues that siting decisions are already being made in a stepwise and participatory way, and that there has been a significant move to increased participation in siting processes in Europe and elsewhere.

These guidelines highlight the importance of interactions between scholars and siting practitioners in providing broader conceptual perspectives and insights into the management of siting processes. The Credo was generated from a workshop that included scholars principally from MIT, Harvard, and Pennsylvania universities and practitioners from the public and private sectors in the United States and Canada. The development of the stepwise-siting approach involved contributions from several scholars in Europe and policy makers associated with the Nuclear Energy Agency (NEA). The report itself stressed the importance of the scholarly literature in the development of this approach by a major international organization. This suggests that the literature is providing important insights and finding its way onto the desks of some siting practitioners.

Instrumental Utility

This discussion supports a contention in the knowledge-utilisation literature that suggests that a major role of scholarly research is an enlightening one, whereby knowledge production can provide useful conceptual insights which can and do, over time, have an impact on the broader orientations and perspective of practitioners (Weiss, 1977; Booth, 1988). Yet, an examination of the siting literature also reveals that it can provide useful instrumental insights in strategic and functional ways. Walt (2005) and Jentleson (1990, 2002) suggest a useful way to examine the usefulness of theory to policy practitioners by reference to its diagnostic, predictive, prescriptive, and evaluative utility. I apply that approach to evaluating the instrumental utility of the siting literature.

Siting theories can assist practitioners in *diagnosis*, or attempting to understand what phenomenon they are facing. For instance, theory can help policy makers understand if the motivations of those opposing projects are based simply on emotional concerns about projects, green ideology, or indeed, simply a desire to extract more benefits out of project developers (Welcomer et al., 2000). While there may be some element of truth in these assertions, siting theories suggest that resistance to projects is based on motivations that are more complex and nuanced (Wolsink, 2007). It also has a lot to do with real and legitimate concerns about the possible and often negative aspects of projects on local communities (including both physical and nonphysical harms), the nature of participatory decision processes involved in siting those projects, and a lack of trust in institutions governing siting processes. Such diagnosis has significant implications for devising an approach to siting and the management of conflict.

Theory can help in understanding and interpreting historical siting experience and guiding practitioners in their responses to the future. Theory provides a broad set of useful diagnostic options for decision makers. International experience reveals that most industrialised nations have generally abandoned DAD approaches in favour of more democratic approaches to locating unwanted facilities. While no single model of siting has emerged, the literature highlights participatory and deliberative responses to siting in democratic nations such as Austria (where strong hierarchical traditions persist), France (where the state has tried to embed itself in local communities), Japan (where the state uses an array of compensatory and other policy tools), and Germany (where cooperative discourse approaches have been attempted) (Lesbirel & Shaw, 2005). While there are variations in the effectiveness of these approaches (they have worked in some cases, but not in others), they do provide practitioners with a useful set of diagnostic possibilities.

Such diagnosis, based on siting theories, can point practitioners in the direction of additional information that is likely to be important in siting conflict management. For instance, competing views of procedural fairness will be an important determinant of siting outcomes. As Linnerooth-Bayer (2005) points out, fairness can be understood in the context of major forms of social organisation (hierarchical, market, and egalitarian). Hierarchical approaches stress authority and procedural rationality, where fairness is settled by administrative determination. Market approaches are distinguished by an emphasis on personal rights, freedoms, and economic rationality where distributive issues are settled by market interactions. Egalitarians reject the unequal social relations contained in both hierarchical and market views of fairness and abhor morally procedures that perpetuate social inequalities such as sited facilities in poor and minority communities on environmental justice grounds. While there is no precise and unambiguous way to measure fairness and equity in siting processes, her conclusions provide a way to help siting practitioners in their search for additional relevant diagnostic information.

Theory can also assist siting practitioners in *prediction*, or anticipation of, conditions, events, and trends that influence the broad environment in which siting occurs. Hunold and Young (1998) highlight the importance of changed levels of cynicism towards the capacity of democracies to promote justice in terms of communicative participation as a key contextual variable influencing facility siting. Kasperson (2005) highlights the importance of a changed social and political context where changed perceptions of risk (including the amplification of risk), trust, and confidence in siting institutions, and equity and environmental justice concerns have created a different context in which the siting process now occurs in democratic nations. Aldrich (2007) stresses the level of social capital in communities as an overall determinant of the locational characteristics of a range of projects in industrialised nations. While siting practitioners might not be able to influence the broader environment in which siting occurs, being able to provide some reasonable predictions will assist them in anticipating how historical, social, and political contexts in which they operate might influence siting processes and strategies.

Such predictions also help stakeholders prevent or manage unwanted, or reinforce wanted, developments in siting processes. Barthe and Mays (2005) provide an excellent analysis to highlight unintended consequences of legislative changes that required more communicative process in facility siting. It shows how such communicative processes, if perceived as not only providing information *to* the public, but also *on* the public, can open up a forum for opposing voices or interests that can derail siting attempts. The siting of the Bayer chemical project in Taiwan during the 1990s shows the importance of anticipating electoral outcomes in siting processes. While the company had made significant efforts to increase local community support for a factory, it was not able to prevent the key leader of the resistance from continuing to politicise the dispute and win a seat in the elections, thereby changing power structures, which ultimately forced the company to abandon the project (Personal Communication, 2002).

Siting processes are not static, and the relationship between predictive theories and real world developments is highly dynamic, making predictions highly problematic. For instance, O'Hare (1977) and O'Hare et al., (1983) highlight the strategic importance of compensation in reducing resistance to projects. Yet, as several scholars have subsequently observed, such predictions can be inaccurate as compensation can inject instabilities in siting processes. It can do this by changing levels of altruism (Frey & Oberholzer-Gee, 1997), increasing concerns that the risks of projects are high and that developers are paying "blood money," especially if mitigation measures have not been employed (Gerrard, 1994; Kasperson, 2005), and changing power relationships and the scope of conflict in siting disputes (Lesbirel, 1998). That may help to understand why many observers argue that economic inducement strategies are ineffective or only effective under limited conditions in managing siting conflicts. While perfect predictions in the social sciences are not possible because of the relationship between those predictions and behaviour, the theory can sensitise practitioners to anticipate the likely consequences from their actions and to account for those in fashioning their approaches to siting.

Siting theories can also facilitate practitioners by providing useful *prescription,* or policy, approaches to achieve desired results. The literature offers a range of examples of siting guidelines that have been based on theory. As discussed earlier, perhaps the best known is the Credo. While it provides conceptual insights, it also provides important instrumental insights that are grouped into three areas. The first relates to goals and objectives, such as instituting a wide participatory process and working to develop trust. The second set concerns appropriate outcomes, such as guaranteeing stringent safety standards will be met, addressing negative aspects of the facility and making the community better. The third relates to steps in the process, such as using a volunteer process, seeking to achieve geographical fairness, and keeping a range of options open at all times (Kunreuther et al., 1993).

Other approaches have suggested siting in stages or steps and have highlighted different aspects of the Credo. Sequential multistage siting processes provide an illustration of a comprehensive stage-based approach. It includes site selection, environmental impact assessment, benefit-cost analysis, mitigation, public hearings, negotiations, and an auctioning process to determine relative compensation requirements (Quah & Tan, 1998, 2002). Stepwise siting involves the development of steps in the siting process that are reversible. The guidelines specify a set of goals, such as having open debate, developing an understanding that the status quo is unacceptable, identifying one more acceptable site, negotiating tailor-made compensation packages, and fully respecting agreements, as crucial in implementing siting solutions that are regarded as legitimate (OECD, 2004). Cooperative discourse approaches entail the three major consecutive steps. The first involves the identification and selection of concerns and evaluative criteria by relevant interests. The second concerns the identification and measurement of impacts, and consequences related to different policy options and establishing expert consensus on these consequences and options. The final step involves conducting a rational discourse with randomly selected citizens as jurors, with citizens and interest group representatives as witnesses, and with citizen panels ultimately deciding among the options (Schneider et al., 2005).

These various siting prescriptions provide useful, strategic insights for siting practitioners. First, they help identify goals and objectives of siting processes, such as equity, efficiency, and liberty, and trade-offs among those goals. Second, they assist in exploring the problems involved in siting conflicts, such as assessing the costs and benefits of projects and dealing with interests that become involved in siting processes. Third, they help in identifying policy instruments, such as inducements, rights, and persuasion, for managing siting conflicts. While there is no strategic approach that is unambiguously favoured among scholars, these prescriptions provide important starting points for practitioners in the strategic crafting of approaches to siting projects.

A final way that siting theories can assist practitioners is in *evaluation,* or specifying, benchmarks for assessing the success or otherwise accomplishing siting objectives. While the concepts of success and failure have often been used loosely by siting scholars, perhaps reflecting their slipperiness as concepts and their normative connotations (Smith, 1989), the literature suggests at least four factors are important when making judgements about success or failure in siting. (1) Siting policies can be judged in terms of design and how they were formulated with success or failure assessed on values and interests. This would entail judging siting policies in terms of whether they were based on appropriate and acceptable values, such as equity, justice, and participation (Linnerooth-Bayer, 2005). (2) Siting policies can also be judged in terms of their execution. For instance, success or failure could be assessed on the efficiency (including social efficiency) and the degree to which siting approaches yielded siting outcomes consistent with their design (Kraft, 2000). (3) Siting policies can be evaluated by the extent to which they were effective in achieving goals and leading to solutions to a societal problem (Kunreuther et al., 1993). For instance, success or failure could be assessed on the extent to which siting policy meets societal needs for those projects without leading to other societal problems, both foreseen and unforeseen. (4) Siting policies can be judged on the overall normative positions of the stakeholders involved. As Lindblom (1959) notes, the only test of a good policy outcome is a consensus that it is good.

CONCLUDING REMARKS

This paper started out with the observation that being a siting practitioner is not an easy task. Scholars rarely do siting, but they do produce knowledge about siting processes and outcomes. It is hoped that this paper has clarified the nature of the knowledge that they produce and how that knowledge can be used by practitioners in conceptual and instrumental ways. It is through the production of ideas and knowledge that scholars can contribute to helping practitioners in their desire and mission to make siting a more manageable task.

The analysis suggests two principal conclusions. (1) The siting literature has evolved into a substantive one. It has used a full range of theoretical and methodological perspectives to explore siting processes and outcomes for a wide range of facilities. It has generated a variety of middle-range explanations that have enhanced our understanding of siting processes and outcomes in rigorous and sophisticated ways. (2) The knowledge that siting scholars have produced is policy-relevant. It offers a range of conceptual and instrumental insights that can assist practitioners in siting process management. Importantly, there is some evidence that these insights have been based on two-way interactions between scholars and practitioners and that they have found expression in observed approaches to siting in many democratic nations.

While the siting literature can offer and has provided useful insights, the challenge will be to build on these achievements by addressing several theoretical and empirical shortcomings in ways that seek to enhance its utility to practitioners.

The first issue relates to the practical utility of contestable theories in the literature. A key feature of the literature is contestability. Indeed, it is the contestable nature of the literature that has allowed it to develop and enhance its explanatory power. Yet, ironically, this process, while extending our knowledge of siting conflicts, might actually act to reduce the practical utility of the literature. Practitioners will be confronted by competing analyses by scholars who are recognised in the field and whose work will be compelling from the perspective they are taking. However, they may be reluctant to use these theories because they may not know how and when to emphasise one theory over another or to how to combine these theoretical perspectives into one. For instance, practitioners will need to know when to emphasise institutional approaches (which focus on rules) over interpretivist approaches (which focus on the structure of meaning). An important area for future research would be to explore how the siting literature can provide policy practitioners with more guidance on the use of contestable explanations of siting conflicts and their management.

The second issue concerns the effectiveness of existing participatory approaches to siting. There are contending positions in the literature. Some authors suggest that existing approaches have not been effective, and there is a need to change institutional structures (Shaw, 2005); others argue that it is too early to test their effectiveness (Kasperson, 2005); still others argue that there is some evidence that such democratic approaches can work effectively to resolve siting conflicts (Rabe et al., 2000). Reconciling these debates will require addressing biases in the literature. With some notable exceptions, it is still heavily biased towards exploring the difficult cases. While these are important, it is equally critical to explore the easier cases. For example, while there continue to be studies that analyse why nuclear projects are abandoned, there are few studies that explore why others have been developed. Currently, there are 33 reactors under construction in a variety of nations such as Canada, China, Finland, France, India, Russia, South Korea, and Japan (IAEA, 2007). Yet, there appear to have been no extensive siting studies that explain these cases collectively. Practitioners will not only want to know about the difficult cases; they will also want to know why other cases were resolved. An important issue for future research will be rectifying these selection biases in the literature and testing such outcomes against the approaches proposed in the literature.

The third issue concerns the contingent nature of middle-range theories and their relative strategic and tactical utility. Siting scholars will often be happy with middle-range theories that produce contingent explanations of the form that an X percent increase in trust will lead to a Y percent increase in the probability of reaching agreement. While such conclusions will be acceptable to scholars and might provide useful strategic insights, practitioners will also wish to know how to address more immediate tactical needs. They will need to know how to overcome the problem of trust. But they will also need to know whether the relationship between trust and probability of acceptance fits the particular circumstances they are confronting or whether that circumstance is an outlier. While some research has pointed to the importance of tactical needs in the development of policy guidelines (Kasperson, 2005), there is considerable scope for future research on making the siting literature more useful to policy makers in terms of their day-to-day tactical needs.

The final issue relates to the utility of conceptual and instrumental insights to siting conflict management more generally. The dominant insights found in the literature relate to North America and Western European experience. There is little analysis of whether such insights will be useful to practitioners in Asia, a region characterized by nations of differing political systems and levels of economic development. While there is some evidence that existing insights are likely to be relevant to nations such as Japan, South Korea, and Taiwan, practitioners in Asia will wish to know if they are applicable to other nations. For instance, an interesting article on siting in China suggests that local governments and host communities can block the establishment of waste facilities as a result of increasing decentralisation of decision-making power as local governments are granted more autonomy in decision-making (Chung et al., 2002). There is also some evidence that siting is becoming an important issue in Vietnam, especially with increased citizen concerns after construction (Cuong, 2003). A critical area of future research relates to the extent to which Western-based siting insights are likely to have applicability in Asia and the extent to which practitioners in different nations can learn from other siting experiences and encourage the transfer of knowledge.

REFERENCES

Aldrich, D. P. (2007). *Site fights: Divisive facilities and civil society in Japan and the West.* Ithaca, NY: Cornell University Press.

Bantjes, R., & Trussler, T. (1999). Feminism and the grass roots: Women and environmentalism in Nova Scotia, 1980–1983. *The Canadian Review of Sociology and Anthropology, 36*(2), 179–180.

Barthe, Y., & Mays, C. (2005). Communication and information: Unanticipated consequences in France's underground laboratory siting process. In S. H. Lesbirel & D. Shaw (Eds.), *Managing conflict in facility siting: An international comparison.* Cheltenham, UK: Edward Elgar.

Been, V. (1994). Locally undesirable land uses in minority neighborhoods: Disproportionate siting or market dynamics? *The Yale Law Journal, 103*(6), 1383–1422.

Been, V., & Gupta, F. (1997). Coming to the nuisance or going to the barrios? A longitudinal analysis of environmental justice claims. *Ecology Law Quarterly, 24*(1).

Birnbaum, R. (2000). Policy scholars are from Venus; Policy makers are from Mars. *Review of Higher Education, 23*(2), 119–132.

Booth, T. A. (1988). *Developing policy research.* Aldershot, UK: Avebury.

Brown, P., & Ferguson, F. I. T. (1995). "Making a big stink": Women's work, women's relationships, and toxic waste activism. *Gender and Society, 9*(2), 145–172.

Bullard, R. (1990). *Dumping in Dixie: Race, class and environment quality.* Boulder, CO: Westview Press.

Caplan, N. S., Morrison, A., & Stambaugh, R. J. (1975). *The use of social science knowledge in policy decisions at the national level: A report to respondents.* Ann Arbor, MI: Institute for Social Research, University of Michigan.

Chung, S. S., Lo, C. W. H., & Poon, C. S. (2002). Factors affecting waste disposal facilities siting in Southern China. *Journal of Environmental Assessment Policy and Management, 4*(2), 241–262.

Cuong, L. D. (2003). *Institutional issues for landfill siting in Viet Nam: Practical recommendations for improvement.* Toronto: University of Toronto.

Dear, M. (1992). Understanding and overcoming the NIMBY syndrome. *Journal of the American Planning Association, 58*(3), 288–300.

Eriksson, J., & Sundelius, B. (2005). Molding minds that form policy: How to make research useful. *International Studies Perspectives, 6*(1), 51–71.

Foucault, M. (1977). *Discipline and punish: The birth of the prison.* London: Allen Lane.

Frey, B. S., & Oberholzer-Gee, F. (1997). The cost of price incentives: An empirical analysis of motivation crowding-out. *The American Economic Review, 87*(4), 746–755.

Gerrard, M. (1994). *Whose backyard, whose risk: Fear and fairness in toxic and nuclear waste siting.* Cambridge, MA: MIT Press.

Haggett, C., & Toke, D. (2006). Crossing the great divide—using multi-method analysis to understand opposition to wind farms. *Public Administration, 84*(1), 103–120.

Halfpenny, P. (1982). *Positivism and sociology: Explaining social life.* London: Allen & Unwin.

Hamilton, J. T. (1993). Politics and social costs: Estimating the impact of collective action on hazardous waste facilities. *The RAND Journal of Economics, 24*(1), 101–125.

Hamilton, J. T. (2005). Environmental equity and the siting of hazardous waste facilities in OECD countries: Evidence and policies. *International Yearbook of Environmental and Resource Economics 2006.*

Hansen, S. (2003). *Pipeline politics: The struggle for control of the Eurasian energy resources.* The Hague: The Clingendael Institute.

Hay, C. (2002). *Political analysis.* Basingstoke, UK: Palgrave.

Hecht, G. (1998). *The radiance of France: Nuclear power and national identity after World War II.* Cambridge, MA: MIT Press.

Hubbard, P. (2005). Accommodating otherness: Anti-asylum centre protest and the maintenance of white privilege. *Transactions of the Institute of British Geographers, 30*(1), 52–65.

Hunold, C., & Young, I. M. (1998). Justice, democracy, and hazardous siting. *Political Studies, 46*(1), 82–95.

Hunter, S., & Leyden, K. M. (1995). Beyond NIMBY: Explaining opposition to hazardous waste facilities. *Policy Studies Journal, 23*(4), 601–619.

IAEA. (2007). International Atomic Energy Agency Pris Database. Retrieved from www.iaea.org/programmes/a2/index.html.

Jenkins-Smith, H., & Kunreuther, H. C. (2005). Mitigation and benefits measures as policy tools for siting potentially hazardous facilities: Determinants of effectiveness and appropriateness. In S. H. Lesbirel & D. Shaw (Eds.), *Managing conflict in facility siting: An international comparison.* Cheltenham, UK: Edward Elgar.

Jentleson, B. W. (1990). Reflections on Praxis and Nexus. *PS: Political Science and Politics, 23*(3), 434–436.

Jentleson, B. W. (2002). The need for Praxis: Bringing policy relevance back in. *International Security, 26*(4), 169–183.

Jopling, D. G. (1974). Plant site evaluation using numerical ratings. *Power Engineering (March),* 56–59.

Kasperson, R. E. (2005). Siting hazardous facilities: Searching for effective institutions and processes. In S. H. Lesbirel & D. Shaw (Eds.), *Managing conflict in facility siting: An international comparison.* Cheltenham, UK: Edward Elgar.

Kraft, M. E. (2000). Policy design and the acceptability of environmental risks: Nuclear waste disposal in Canada and the United States. *Policy Studies Journal, 28*(1), 206–218.

Kruize, H., Driessen, P. P. J., Glasbergen, P., & van Egmond, K. N. D. (2007). Environmental equity and the role of public policy: Experiences in the Rijnmond region. *Environmental Management, 40*(4), 578–595.

Kunreuther, H., Fitzgerald, K., & Aarts, T. D. (1993). Siting noxious facilities: A test of the facility siting credo. *Risk Analysis, 13*(3), 301–318.

Kunreuther, H. C. (1995). *The dilemma of siting a high-level nuclear waste repository.* Massachusetts: Springer.

Lane, R. (1990). Concrete theory: An emerging political method. *The American Political Science Review, 84*(3), 927–940.

Lesbirel, S. H. (1998). *NIMBY politics in Japan: Energy siting and the management of environmental conflict*. Ithaca, NY: Cornell University Press.

Lesbirel, S. H., & Shaw, D. (2005). *Managing conflict in facility siting: An international comparison*. Cheltenham, UK: Edward Elgar.

Lindblom, C. E. (1959). The science of "muddling through." *Public Administration Review, 19*(2), 79–88.

Linnerooth-Bayer, J. (2005). Fair strategies for siting hazardous waste facilities. In S. H. Lesbirel & D. Shaw (Eds.), *Managing conflict in facility siting: An international comparison*. Cheltenham, UK: Edward Elgar.

Lober, D. J. (1995). Why protest?: Public behavioral and attitudinal response to siting a waste disposal facility. *Policy Studies Journal, 23*(3), 499–518.

Loefstedt, R. E. (1996). Fairness across borders: The Barsebaeck nuclear power plant. *Risk-Health Safety & Environment, 7*(2), 135–144.

Marsh, D., & Stoker, G. (Eds.). (2002). *Theories and methods in political science* (2nd ed.). Basingstoke, UK: Palgrave Macmillan.

McAvoy, G. E. (1994). State autonomy & democratic accountability: The politics of hazardous waste policy. *Polity, 26*(4), 699–728.

Morell, D., & Magorian, C. (1982). *Siting hazardous waste facilities: Local opposition and the myth of preemption*. Cambridge: Ballinger.

Munton, D. (1996). *Hazardous waste siting and democratic choice*. Washington, DC: Georgetown University Press.

O'Hare, M. (1977). Not on my block you don't: Facility siting and the importance of compensation. *Public Policy, 25*, 407–458.

O'Hare, M., Bacow, L. S., & Sanderson, D. (1983). *Facility siting and public opposition*. New York: Van Nostrand Reinhold.

Organization for Economic Development and Cooperation (OECD) (2004). *Stepwise approach to decision making for long-term radioactive waste management*. Retrieved from www.nea.fr/rwm/reports/2004/nea4429-stepwise.pdf

Pescatore, C., & Vári, A. (2006). Stepwise approach to the long-term management of radioactive waste. *Journal of Risk Research, 9*(1), 13–40.

Quah, E., & Tan, K. C. (1998). The siting problem of NIMBY facilities: Cost-benefit analysis and auction mechanisms. *Environment and Planning C. Government & Policy, 16*(3), 255–264.

Quah, E., & Tan, K. C. (2002). *Siting environmentally unwanted facilities: Risks, trade-offs, and choices*. Cheltenham, UK: Edward Elgar.

Rabe, B. G. (1994). *Beyond NIMBY: Hazardous waste siting in Canada and the United States*. Washington, DC: Brookings Institution.

Rabe, B. G., Becker, J., & Levine, R. (2000). Beyond siting: Implementing voluntary hazardous waste siting agreements in Canada. *American Review of Canadian Studies, 30*(4), 479–496.

Sakai, T. (2005). A normative theory for the NIMBY problem. Mimeo, Yokohama City University.

Sayer, R. A. (1992). *Method in social science: A realist approach* (2nd ed.). London and New York: Routledge.

Schneider, E., Oppermann, B., & Renn, O. (2005). Implementing structured participation for regional level waste management planning. In S. H. Lesbirel & D. Shaw (Eds.), *Managing conflict in facility siting: An international comparison* (xii, 220). Cheltenham, UK: Edward Elgar.

Shaw, D. (2005). Visions of the future for facility siting. In S. H. Lesbirel & D. Shaw (Eds.), *Managing conflict in facility siting: An international comparison* (xii, 220). Cheltenham, UK: Edward Elgar.

Slovic, P. (2000). *The perception of risk.* London: Earthscan.

Smith, T. B. (1989). The analysis of policy failure: A three-dimensional framework. *Indian Journal of Public Administration, 35*(1), 1–15.

Stone, D., Maxwell, S., & Keating, M. (2001). Bridging research and policy. Paper presented at an international workshop funded by the UK Department for International Development.

Vári, A., Reagan-Cirincione, P., & Mumpower, J. L. (1994). *LLRW disposal siting: Success and failure in six countries.* Dordrecht, Netherlands: Kluwer Academic Publishers.

Walt, S. (2005). The relationship between theory and policy in international relations. *Annual Review of Political Science, 8,* 23–48.

Weingart, J. (2001). *Waste is a terrible thing to mind: Risk, radiation, and distrust of government.* Princeton: Center for Analysis of Public Issues.

Weiss, C. H. (1977). *Using social research in public policy making.* Lexington, MA: Lexington Books.

Welcomer, S., Gioia, D., & Kilduff, M. (2000). Resisting the discourse of modernity: Rationality versus emotion in hazardous waste siting. *Human Relations, 53*(9), 1175–1205.

Wilson, E. J. (2000). How social science can help policymakers: The relevance of theory. In M. Nincic & J. Lepgold (Eds.), *Being useful: Policy relevance and international relations theory* (109–128). Ann Arbor, MI: University of Michigan Press.

Wolsink, M. (2007). Wind power implementation: The nature of public attitudes: Equity and fairness instead of "backyard motives." *Renewable and Sustainable Energy Reviews, 11*(6), 1188–1207.

3

Procedures for Dealing with Transboundary Risks in Siting Noxious Facilities[1]

Howard Kunreuther

INTRODUCTION

This chapter focuses on ways to better manage the risks associated with facilities that pose health and environmental consequences that can affect a wider population than the area in which the facility is located. In other words, these proposed facilities pose *transboundary risks* due to the negative externalities that they would create with respect to the ecosystem and neighboring districts. A strategy for siting such facilities should focus on the concerns of the relevant stakeholders and the nature of the risk. The next section describes the nature of the problem from the perspective of the different stakeholders or interested parties affected by the risk. I then propose a framework for evaluating alternative siting strategies by focusing on where to locate an incinerator as an illustrative example. The chapter then examines how a siting authority can deal with transboundary issues and the role that mitigation measures and compensation can play in gaining consensus by the public and other key stakeholders as to whether the facility should be approved. The concluding section

[1] This chapter is based on studies undertaken with my colleagues involved in the dialogue associated with siting a high-level radioactive waste repository in Nevada, as well as related papers on siting locally unwanted land use (LULU) facilities. The ideas also reflect many discussions on siting issues over the years with Doug Easterling, Jim Flynn, Robin Gregory, Hank Jenkins-Smith, Roger Kasperson, S. Hayden Lesbirel, Joanne Linnerooth-Bayer, Michael O'Hare, Daigee Shaw, Paul Slovic and Larry Susskind. Support from NSF Grant No. CMS-0527598 and the Wharton Risk Management and Decision Processes Center is gratefully acknowledged.

elaborates on the elements of the Facility Siting Credo that may be helpful in finding a site for a facility with transboundary risks and raises a set of issues and questions for implementing this approach in areas such as Hong Kong that face these problems.

NATURE OF THE PROBLEM

The nature of the transboundary risk problem from the perspective of the affected districts can be stated by using the following simple example. District Y is planning to site a facility that not only affects its own residents but also those of District Z. There may be benefits to both Y and Z from having the facility, but it is likely that Y gains considerably more than Z does either through tax revenues and/or employment opportunities for its residents. There are three interrelated questions that need to be addressed from the perspective of both districts:

(1) What actions should District Y take with respect to mitigating its risks, recognizing that the negative impacts (e.g., pollution) may extend beyond its own boundaries?
(2) Is there a role that a siting authority can play in managing these transboundary risks?
(3) What role can compensation or benefit-sharing by District Y play in satisfying the concerns of District Z?

There are a number of parties who are affected by the facility that need to be considered in the siting process:

The developer, who is interested in constructing the facility: In many countries the developer is synonymous with a government organization. For example, in Hungary and Slovakia, the water management authority is responsible for developing hydroelectric power plants to manage their country's energy needs. Government agencies in Lithuania and Sweden are responsible for providing nuclear power as a source of energy.

The affected public, who both benefit from the facility and are affected by the risk. Residents in districts near the facility may be more adversely impacted by it than those some distance away. In Hong Kong, the local community has been concerned with the negative impact of a proposed central slaughterhouse over the risks of bird flu and the decline in property values.

Public interest groups, who have their own agenda regarding future development projects: For example, "green" groups in Hong Kong have been concerned with the impact of a proposed liquefied natural gas-receiving terminal on the local ecology and the impact of a proposed waste incinerator on the environment.

Two questions that arise when dealing with transboundary risks and stakeholder groups are: (1) What role should the public and environmental groups have in making decisions regarding the siting and operation of certain facilities? (2) How does one create trust in the process of siting facilities and managing them when there is great uncertainty associated with risks?

A FRAMEWORK FOR ANALYZING THE TRANSBOUNDARY PROBLEM

Consider the challenges facing a private firm or developer who is trying to find a site for a solid or hazardous waste facility. As a concrete example, suppose that the facility of interest is an incinerator. District Y has expressed an interest in hosting the incinerator right near its political boundary in a part of town that has relatively few homes and businesses. Residents in District Z are also subject to health and environmental risks from the facility.

A voluntary siting process has been proposed whereby all the eligible residents in Y can vote on a referendum as to whether an incinerator should be located in their backyard. Suppose that if a certain proportion (e.g., two-thirds) of Y's residents support the facility, then it will be deemed approved and construction will begin. Those residing in Z have no official vote on whether the incinerator should be located in Y, but they can publicly protest the facility in order to encourage residents in Y to vote against it.

Each resident j in District Y will determine whether to vote in favor of a particular facility by considering the benefits to him or her (B_j) and the perceived risks associated with the facility. The benefits can be direct compensation to an individual, such as a reduction in property taxes, or it can take the form of community-wide or regional improvements, such as additional health-related services, higher salaries to attract more and better teachers for the schools, and/or new recreational facilities. A benefits package may also contain contingent arrangements such as guarantees against property-value declines due to the facility, and reimbursement for any health and/or environmental impacts from the facility.

The risks associated with the incinerator for each individual j in District Y are characterized by a perceived probability (p_j) that some type of damage (D_j) will occur to him or her. These risks can be mitigated (but not necessarily eliminated) through enforcement of safety standards and regulations. If the benefits package is attractive enough and/or the perceived risks associated with the facility are sufficiently small to resident j in District Y, then he or she will vote "Yes" to constructing the incinerator.

The developer has *no* economic incentive to provide residents in District Z with a benefit package or to reduce the risks facing this group. From the developer's perspective, the only votes that count are the ones from District Y. Thus it is conceivable that the majority of the residents in District Z may face certain risks from the new facility for which they will not be sufficiently compensated by the developer, so that they would disapprove of the facility even though it was authorized by those living in Y.

This example illustrates the divergence between private and social costs due to transboundary risks. The private costs to the developer revolve only around residents in Y, while the social costs include the impact on individuals in both Y and Z. Unless some steps are taken to protect District Z against possible economic, health, and safety losses, the above voluntary siting process can prove costly when the benefits and costs to residents in both districts are taken into account.

Transboundary risks are a form of externalities that are normally associated with public goods or bads. Thus, a hazardous waste facility poses risks of different degrees to all individuals within a certain radius of the site, and there is little that a person can do to alleviate this risk once a facility is built, short of moving out of the area. Individuals can engage in collective action to lobby against having a facility in the first place, but this involves costs to them that they may not be prepared to incur. Hamilton (1993) has shown that private firms will want to locate facilities in communities or regions that generate the least political opposition, and he provides empirical evidence that the host communities will not necessarily be the ones that generate the lowest externalities.

Importance of International Siting Authority

The presence of externalities suggests a clear role for a siting authority or government agency to play an active role in the siting process. More specifically, such a group would need to impose strict mitigation measures and standards that reduce the risks to both the host community and its affected neighbors before any developer or firm engages in the search for a site. It would also specify who is liable in case an accident occurs and what the appropriate compensation would have to be.

Role of Well-Specified Standards. To adequately reflect the concerns of all the affected residents, the siting authority would need jurisdiction over a region that encompasses both the host district as well as those areas subject to the transboundary risks from a facility. Furthermore, it would need to be empowered by a governmental body that was concerned with the welfare of a wide area rather than the narrower interests of citizens from one jurisdiction.

There is an additional reason for imposing strict safety standards by a public authority or governmental authority before the search for a site begins. The standards are likely to reduce conflicts that may otherwise emerge between a developer who relies on scientific experts for characterizing risks and the residents in the community who have their own perceptions of the risks. While the experts normally measure risks in quantitative terms (e.g., the probability and the anticipated consequences of an accident), the public takes other factors such as dread, unfamiliarity with the technology, and catastrophic potential into account when evaluating their concerns (Slovic, 1987). Residents in Z who oppose the facility can feed into the fears of those residing in Y, encouraging them to vote against the proposal.

Evidence on how the public's perceptions of risk differ from the scientists' views is illustrated by two empirical studies. One study showed that the amount that a layperson was willing to pay for risk reductions is influenced by his or her degree of dread and the severity of risks, such as hazardous waste and sulfur air pollution, where there is considerable scientific uncertainty in the degree of risk exposure and their potential effects. Scientists do not consider factors such as dread or lack of familiarity to be relevant in characterizing the degree of risk from a particular activity (McDaniels, Kamlet, & Fischer, 1992). Another study of laypersons and toxicologists revealed large differences between these two groups in their assessment of chemical risks (Kraus et al., 1992).

Role of Monitoring and Control Procedures. In addition to imposing standards and regulations at the time the facility is sited to deal with risk perceptions of both the experts and the public, the public authority needs to undertake monitoring and control procedures at regular intervals to assess the performance of the facility once it is in place. One proposal that may convince the affected public that they will be protected against risks to themselves and future generations is to form a committee of local residents that is granted special oversight powers, including the power to suspend operations at the facility if the prescribed standards are not adhered to.

Use of Compensation or Benefit Sharing

For residents to support a facility, the benefits associated with having it must be greater than the benefits of maintaining the status quo. One way to satisfy this condition is to provide compensation to communities that site the facility, as well as those that are nearby. However, compensation will be viewed as a bribe unless the affected groups feel that the facility satisfies rigorous safety standards that will be well enforced. Six types of compensation have been identified by Gregory et al., (1991) for facilitating siting decisions.

Direct monetary payments. This is the most common form of compensation and can take the form of guaranteed annual payments on the waste that is stored. In Charles City, Virginia, the developer of a landfill collected a fee that it paid to the city, amounting to about $1 million in revenues annually. This lowered property taxes and allowed for the rebuilding of the city's ailing school system (O'Hare et al., 1983).

In-kind awards. These take the form of grants to communities or regions for improving health care facilities, housing, education, or other services that enhance the citizens' well-being and reduce risks they face. In Swan Hills, Alberta, subsidized housing was provided for 35 housing units in conjunction with the siting of a hazardous waste facility (Rabe, 1991). In Charles City, the operator collected the county's garbage free of charge. This was also the case in Grandview, Idaho where Wes-Con provided free garbage pickup to residents as part of a package associated with siting a waste-disposal facility (O'Hare et al., 1983).

Contingency funds. These are used to cover losses from an accident or other adverse effects of the facility. For example, trust funds could be established to cover the damages and health-related costs to victims of an accident.

Property value guarantees. These protect residents of the host community and surrounding areas against any decline in the resale value of their home due to the location of the facility. This type of compensation was offered by Champion International Corp. as part of a program for siting an industrial landfill. The company monitored the changes in the sale prices of homes in the county over a ten-year period and paid residents if there were any adverse changes in the property value due to the presence of a landfill (Ewing, 1990).

Benefit assurances. These guarantee direct or indirect employment for community members, either during construction of the facility or during its operation phase. These types of benefits have positive externalities that will make the facility more attractive to neighboring communities as well as the host site. The hazardous waste treatment centre in Swan Hills, Alberta, promised new jobs and convinced town leaders that other developments, such as a new hospital, would now be feasible.

Economic goodwill. This refers to contributions to local organizations and expenditures for projects that are important to the community and the surrounding area. The private corporation responsible for the Swan Hills hazardous waste facility planted 400 trees for town beautification, provided $65,000 to support local activities including golf course development, and made charitable contributions such as sponsoring a hockey school and donating a bear-skin rug to the town council chambers (Rabe, 1991).

Empirical Evidence on Compensation[2]

A number of surveys have been completed on the attitude of residents toward a public authority that would impose well-specified standards that are monitored and the role compensation can play in facilitating the siting process. This section briefly summarizes some of the key findings for several types of facilities.

Two surveys of particular relevance to this paper are by Bacot, Bowen, and Fitzgerald (BBF) (1994) and by Jenkins-Smith and Kunreuther (JK) (2001), each of which asked respondents to consider compensation in the context of a landfill for municipal waste.[3] Respondents were first asked to indicate whether they would "accept" the construction of a landfill at a nearby site with no mention of benefits.[4] As shown in Table 1, a local landfill was acceptable to 30 percent of the

---

[2] For more details on the empirical findings discussed in this section see Kunreuther and Easterling (1992).

[3] Bacot, Bowen, and Fitzgerald (1994) surveyed 844 Tennessee residents in 1989. Jenkins-Smith and Kunreuther (2001) surveyed 1,200 U.S. households in 1993. This latter sample was split into eight experimental conditions, defined by the type of facility being considered (municipal waste landfill, hazardous waste incinerator, medium-security prison, or high-level nuclear waste repository) and by the order in which the respondent was presented with various compensation and mitigation measures. The effect of economic benefits for any given facility is assessed here with a sub-sample of 150, where benefits are offered as the first measure.

[4] In the Bacot, Bowen, and Fitzgerald (1994) survey, respondents were told that the landfill was proposed for a site five miles from their home. Acceptance was gauged by a voting question. Jenkins-Smith and Kunreuther (2001) experimentally manipulated the supposed distance to the landfill (either one or ten miles away). Respondents indicated how acceptable such a facility would be. We have coded a respondent as "accepting" the facility if he or she gave a response of either "acceptable" or "completely acceptable."

BBF sample and to 25 percent of the JK sample when compensation was *not* included. However, in both cases, the rate of acceptance approximately doubled with the introduction of compensation. In the JK survey, the form of the benefits was left vague ("economic benefits provided to residents within 50 miles of the facility"), whereas BBF provided respondents with specific forms of compensation—rebates on property taxes, state money for schools, and state money for road improvements. Tax rebates produced the greatest level of acceptance (63 percent).

Table 1. Effect of Compensation Measures in Increasing Acceptance of Facilities (Percentage agreeing to accept facility)

	Municipal waste landfill-study: study 1[a]	**Municipal waste landfill-study: study 2[b]**	**Incinerator[b]**	**Prison[b]**
Acceptance **without** incentives	30%	25%	15%	39%
Acceptance **with** economic benefits		49%	30%	52%
Acceptance **with** economic benefits: Rebates on property tax	63%			
Acceptance **with** economic benefits: State money for schools	62%			
Acceptance **with** economic benefits: State money for roads	56%			

[a] Bacot et al., (1994). Sample of 844 Tennessee residents. The 30 percent figure for acceptance without incentives was derived from the reported result that 70 percent opposed the landfill; 30 percent is an upper bound on the actual figure.

[b] Jenkins-Smith and Kunreuther (2001). Total sample of 1,200 U.S. residents. Each condition has n = 150.

The JK survey also investigated the impact of compensation on acceptance in the case where the facility being sited was a hazardous waste incinerator and a medium-security prison. These two facilities differed markedly in the absolute level of acceptability (15 percent versus 39 percent in the no-compensation case); however, the introduction of benefits produced similar levels of increased acceptance (15 percentage points for the incinerator, 13 percentage points for the prison) as shown in the last two columns of Table 1. From these data, one might conclude that economic benefits have a substantial impact on public sentiment toward noxious facilities, although they fail to convince everyone that the facility should be built.

Radioactive waste repositories. The positive impact of compensation on public acceptance is *not* replicated when the facility to be sited is a radioactive waste repository. This conclusion is supported by the five separate studies reported in Table 2: Carnes et al. (1983); Kunreuther et al. (1990); Dunlap and Baxter (1988); Herzik (1993); and Jenkins-Smith and Kunreuther (2001).[5] The different samples varied somewhat in their baseline willingness to accept a "local" high-level nuclear waste repository (HLNW) with the greatest level of acceptance (60 percent) occurring among Dunlap and Baxter's (1988) sample of residents living near Hanford, Washington. However, in none of the surveys did the introduction of benefits produce a major increase in acceptance. A small increase occurred in the Carnes et al. (1983) and JK surveys. However, in the three surveys which offered 20 years of generous tax rebates, there was no evidence of increased acceptance.

---

[5] Carnes et al. (1983) surveyed 420 Wisconsin residents in 1980 on whether they "favored" the siting of a "nuclear waste repository in their community." Kunreuther et al. (1990) conducted a survey of 1,001 Nevada residents in March 1987. Approximately half of these persons (n = 498) were asked about their willingness to vote for an HLNW repository at Yucca Mountain, with and without rebates; the other half was asked about their willingness to pay to have the repository located somewhere else (see Kunreuther and Easterling [1992] for results). Herzik (1993) used a similar rebate question in a 1993 survey of 1,212 Nevada residents. Dunlap and Baxter (1988) also used this sort of question in a survey of 658 residents of Franklin and Benton Counties in Washington State. Respondents indicted their willingness to vote for an HLNW repository at Hanford (which was then still in contention). Jenkins-Smith and Kunreuther (2001) asked 150 U.S. residents "how acceptable" an HLNW repository would be if it were located either ten miles or fifty miles from their home (distance was varied experimentally).

Table 2. Limited Effectiveness of Compensation: The Case of Nuclear Waste Repositories

	Study 1[a]	Study 2[b]	Study 3[c]	Study 4[d]	Study 5[e]	Study 6[f]
Acceptance **without** incentives	22%	10%	27%	24%	60%	5%
Acceptance **with** economic benefits						25%
"substantial payments"	26%					
"economic benefits"		13%				
$1,000/yr for 20 yrs			26%	23%		
$3,000/yr for 20 yrs			30%			
$5,000/yr for 20 yrs			30%			
$100–900/yr for 20 yrs					51%	

[a] Carnes et al. (1983) 1980 survey of 420 Wisconsin residents
[b] Jenkins-Smith and Kunreuther (2001); total sample of 1,200 U.S. residents; each condition has n = 150
[c] Kunreuther et al. (1990); 1987 survey of 1,001 Nevada residents (n = 498 answered compensation questions)
[d] Herzik (1993); 1993 survey of 1,212 Nevada residents
[e] Dunlap and Baxter (1988); 1987 survey of 658 persons living near Hanford, Washington
[f] Frey et al. (1996); 1993 survey of 305 persons living in Wolfenschiessen, Switzerland

In Kunreuther et al. (1990), 27 percent of the sample voted to put a repository at Yucca Mountain in response to a question that did not mention compensation, compared to 29 percent when rebates were offered.[6] This

[6] The referendum question was worded, "If a vote were held today on building a permanent repository, where would you vote to locate the repository?" Respondents were presented with four choices: Yucca Mountain, Hanford, Deaf Smith, and "none of the above." The following question was used to assess a respondent's willingness to accept a repository with compensation:

> Suppose after thorough study, the Federal government decided to put a high-level nuclear waste repository at Yucca Mountain in Nevada. This repository

difference was not statistically significant, $\chi^2(3)$. In addition, there was no significant difference in acceptance across the three dollar amounts: $1,000 per year (26 percent), $3,000 per year (30 percent), and $5,000 per year (30 percent).[7]

The contrast between radioactive waste repositories and other noxious facilities on the effectiveness of compensation is remarkable. *Threat to future generations* is a strong determinant of voting behaviour in the case of HLNW repositories (Kunreuther et al., 1990). If a person believes that a repository will pose serious risks to future generations, rebates are unlikely to win his or her acceptance of a repository. This resistance to rebates is illustrated in Figure 1, which shows the proportion of respondents in the 1987 Nevada survey who favour a repository at Yucca Mountain (with rebates) as a function of perceived risk to self and risk to future generations. This figure shows that the majority of respondents reject rebates if *either* the perceived risk to self is high or the risk to future generations is deemed serious. Among respondents with both beliefs, only eight percent vote in favor of the repository when rebates are offered.

---

would be built according to Federal safety standards. Suppose also that you could receive a [either $1,000/$3,000/$5,000] rebate or credit on your Federal income taxes each year for 20 years. Would you vote to locate the repository at Yucca Mountain?

[7] The effect of dollar amount was nonsignificant, regardless of whether the dependent variable was vote in favor of the repository, or change in acceptability of the repository. In the latter case, respondents were classified into one of three categories: (1) Rebate had no effect on voting response; (2) Rebate made repository more acceptable; or (3) Rebate made repository less acceptable. The effect of rebate level was then assessed by testing whether the distribution of this change variable differed across the three dollar amounts. This yielded a $\chi^2(4)$ of 4.16 ($p > 0.3$).

Very similar responses to compensation were obtained in a 1987 national survey conducted by the same authors (Kunreuther et al., 1990). Here, respondents were asked whether they would accept a repository 50 or 100 miles away in return for rebates of between $1,000 and $5,000 (distance and dollar amount were varied experimentally). Overall, 28.7 percent of the sample responded positively to the compensation offer, compared to 28.9 percent of the Nevada sample.

Figure 1. Approval of Yucca Mountain by Perceived Risk to Self and Risk to Future Generations

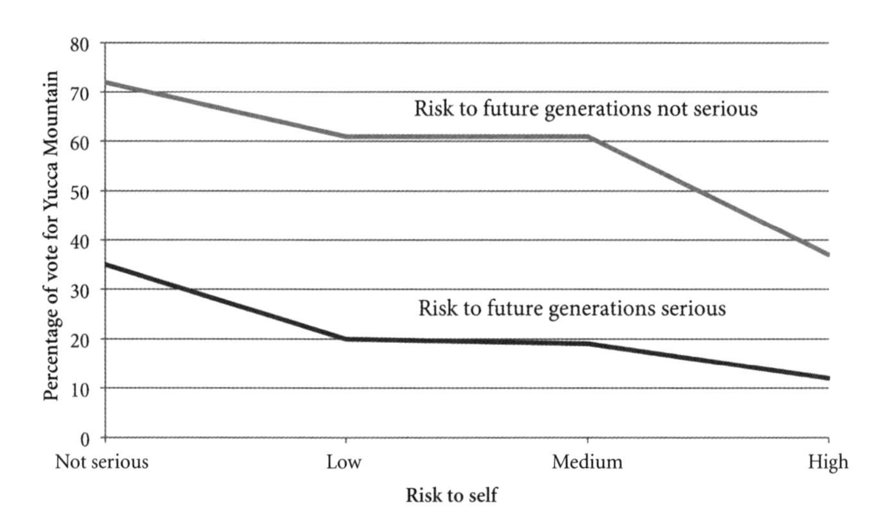

The data in Figure 1 cast doubt on one of the assumptions that underlie most compensation strategies: Compensation will succeed in gaining a person's acceptance of a facility if that person believes he or she will be better off with the facility than without it. Compensation is likely to be rejected whenever a person believes that the proposed facility is somehow *illegitimate* (i.e., should not be built on ethical or moral grounds). This conclusion is supported by McClelland and Schulze (1991). In their study, subjects were given a Norfolk pine at the outset of the task and were asked to indicate the price at which they would sell the pine back to the experimenter. In the condition where subjects were not told anything regarding the fate of the plant, the average asking price was $8. However, among subjects who were told that the plant would be destroyed at the end of the experiment, the average asking price was $18 and a number of subjects reported an asking price that they knew was higher than the experimenter would accept.

Individuals who consider the proposed HLNW repository to be illegitimate will similarly be inclined to reject offers of compensation. The facility might be viewed as illegitimate because of a perceived inequity in the distribution of risks across generations or because of beliefs about the potential of the facility to contaminate the planet (Easterling & Kunreuther, 1995). Monetary payments are inherently unable to offset these objections. For example, a rebate package paid out over 20 years rewards the current generation for accepting the repository, but imposes uncompensated costs on future generations.

Empirical Evidence on Well-Enforced Standards[8]

Requiring stringent standards that address public concerns with the risk implies a greater likelihood that a positive vote will be forthcoming if a siting referendum were instituted. In the JK survey, individuals were asked about their attitudes toward alternative risk reduction measures as a condition for siting four types of facilities: a prison, a landfill, an incinerator, or a high level radioactive waste repository. After the respondents stated their degree of acceptability of one of the four facilities, they were given a series of questions to determine whether one or more of the following measures would cause them to change their stated opinion:

- An independent agency approved by the local government will perform regular inspections to insure that the facility is meeting all federal and state regulations (INSPECT);
- The facility will not be built until local elected officials have approved the design (APPROVE);
- Local elected officials will have the authority to shut down the facility if they detect any problems (SHUTDOWN); and
- Economic benefits were to be provided to residents living within 50 miles of the facility (BENEFITS).

The sample was randomly divided into two equal-sized groups. *GROUP 1* respondents were given the measures, in an additive fashion, in the following order; INSPECT, APPROVE, SHUTDOWN, and BENEFITS. *GROUP 2* respondents began with BENEFITS, followed by INSPECT, APPROVE, and SHUTDOWN, in that order. The reason for the different orderings was to test whether the sequence of measures affected their acceptance of the facility.

Tables 3a and 3b present the percentage of respondents who would support the facility for GROUP 1 and GROUP 2 under five scenarios. The scenarios consist of the combinations of mitigation and compensation packages that were given to the survey respondents. The tables also show whether the change in the percentage of respondents who accept the facility after each of the sequential steps in the mitigation scenarios was statistically significant.

---

[8] This subsection is based on material in Jenkins-Smith and Kunreuther (2001). More details can be found in the paper.

Table 3a. Percentage Completely or Mostly Accepting Facilities Based on Combinations of Safety and Economic Benefits Measures (Group 1 Respondents: Benefits Offered Last)

Sequence of measures	Prison	Landfill	Incinerator	Repository
No Measures	30.5%	18.1%	14.5%	12.4%
INSPECT	54.9%***	53.7%***	42.1%***	31%***
APPROVE	51.8%	46.3%	36.8%	26.2%*
SHUTDOWN	59.8%	65.8%***	55.4%***	42.8%***
BENEFITS	62.8%	56.4%	42.8%***	31.7%

Statistical significance of change from cell above: *** = <0.001; ** = <0.01; * = <0.05

Table 3b. Percentage Completely or Mostly Accepting Facilities Based on Combinations of Safety and Economic Benefits Measures (Group 2 Respondents: Benefits Offered First)

Sequence of measures	Prison	Landfill	Incinerator	Repository
No Measures	39%	25.2%	14.5%	10.2%
BENEFITS	52.1%***	49.0%***	30.3%***	13.4%
INSPECT	65.1%**	67.7%***	47.4%***	30.0%***
APPROVE	56.8%**	57.4%**	39.5%*	25.5%*
SHUTDOWN	56.2%	75.5%***	52.0%***	42.0%***

Statistical significance of ch.ange from cell above: *** = <0.001; ** = <0.01; * = <0.05

Focusing on the specific measures, INSPECT and SHUTDOWN have strong positive effects on acceptance of the facility for both GROUPS 1 and 2. In other words, whether BENEFITS are offered first or last has no impact on the effects of these two measures. Tables 3a and 3b also reveal that the APPROVAL of design by local officials consistently has a *negative* impact on the percentage of respondents who support the siting of these four types of facilities.[9] This finding suggests that, from the perspective of the public, the oversight of independent inspectors and the power to shut down facilities

[9] For GROUP 1, the APPROVE measure is statistically significant only for the repository. However, for GROUP 2, it is statistically significant (in reducing acceptability) for all four facilities.

should be left in the hands of local officials. On the other hand the technical issue of the approval of the facility design should not be delegated to government officials.[10]

Two explanations for this differentiation of public views of the roles of local officials have been offered. One, consistent with the findings of Jenkins-Smith and Stewart (1998), suggests that while local officials are trusted to express the residents' interests, they are not seen as sufficiently competent to directly oversee a complex hazardous materials management program. O'Connor et al. (1994) offer an alternative explanation: Local officials may be seen as too susceptible to the influence of the facility operator, and therefore, cannot be trusted with decisions about facility design. Either way, program designs that seek public acceptance for hazardous facilities must account for the differentiation in public expectations of different kinds of public officials.

Turning to the BENEFIT measure, it makes a big difference whether it is offered first or last. For the landfill, incinerator, and repository, the final percentage of supporters is at least 10 percent higher when economic benefits are offered first (Table 3b) rather than last (Table 3a). Apparently, when the facility is perceived to be risky, providing economic benefits *after* safety measures have been instituted has a negative influence on the percentage of individuals supporting the facility. This finding implies that when economic benefits are offered first, they are more likely to be perceived by the respondents as compensation for the increased risk from hosting a facility. When these benefits are offered *after* safety measures have been addressed, they are perceived by some as a bribe for taking the facility.

This finding is a puzzling one, and requires further research. Indeed, it seems counterintuitive in the light of Kasperson's (1999) argument that compensation is only ethically justifiable after safety measures have been addressed. Our findings suggest to us that, in the relatively untrusting times in which we live, the introduction of benefits after the safety issue has been addressed leads many of those affected to suspect that the facility is even more dangerous than they were initially led to believe. After all, if the facility is safe, why should those living nearby need to be plied with goodies? Hence, support for the facility is eroded by providing benefits as an apparent afterthought.

[10] More research is needed to determine whom people would trust to judge technical issues with respect to facility design. For work on this question, see Jenkins-Smith and Silva (1998).

A SITING PROCEDURE FOR DEALING WITH TRANSBOUNDARY RISKS

At a National Workshop on Facility Siting in 1990, a group of practitioners and researchers developed a set of guidelines for siting noxious and/or hazardous facilities. These guidelines, which were formalized in a Facility Siting Credo, focused on developing a workable and fair procedure for locating a facility as well as an outcome that satisfied distributional (equity) and benefit-cost (efficiency) considerations.

A study of 29 siting cases, both successful and unsuccessful, across the United States and Canada, confirmed the importance of two features of the process of finding a community that agrees to host a facility: having a broad-based public participation process, and the perception by host community residents that the facility was the best solution to their waste problem (Kunreuther et al., 1993). Both elements should be considered in designing a siting process.

The Facility Siting Credo also emphasized the desirability of a voluntary siting process but did *not* explicitly take into account the presence of transboundary risks. This section proposes a two-stage siting process that explicitly addresses the issues of transboundary risks while addressing the concerns of equity and efficiency through compensation. The implicit assumption is made that a new facility is viewed as socially desirable. However, if a volunteer site cannot be found, then the status quo will be maintained rather than forcing a community or region to site a facility. The key questions are what type of facility to construct and where it should be located.

Stage 1: Screen Appropriate Sites and Specify Standards

In this first stage of the process, the Public Siting Authority (PSA) determines a set of sites that meet prespecified technical criteria. At the same time, the PSA specifies a set of safety standards that a proposed facility will have to meet. The PSA can be based at the local, regional, or national level depending on the nature of the transboundary risks and the candidate areas for the facility. The PSA could consist of representatives from more than one country if the facility poses transnational risks.

The screening and standard-setting process should take into account both the risks of the facility to the host community as well as the expected impact it will have on the surrounding areas. If there are transportation risks associated with shipping the material from sources to their final destination, then this factor should play a role in determining what sites are suitable candidates. If the facility has the possibility of causing air pollution to neighboring areas, then this risk needs to be considered when setting specific performance standards for the facility.

One issue that should be addressed in screening acceptable sites is whether to exclude certain communities or regions on equity or fairness grounds. There are two extreme views normally taken with respect to this question. If a voluntary siting process is to be used, then one can argue that any community can decide for itself whether it wants the facility. On the other hand, suppose that most taxpayers feel that low-income areas which already have noxious facilities should be excluded from consideration. Then a siting map should be drawn that excludes these places from being considered, even though they may be technically suitable areas.

Stage 2: Engage in a Voluntary Siting Process

The proposed procedure for finding a site is a voluntary one based on a procedure that was successfully used in Alberta. Fourteen communities were initially interested in hosting a proposed hazardous waste facility. Nine of these were subsequently eliminated either on environmental grounds or because of strong public opposition. Of the remaining five, Swan Hills presented a proposal (including benefits) that best met the needs of the developer (McGlennon, 1983).

A similar procedure was used in Illinois in an attempt to find a home for a low-level radioactive waste repository (English, 1992) and by the Nuclear Waste Negotiator in an attempt to find a state or Indian tribe willing to host an MRS facility for the temporary storage of spent nuclear fuel (Office of Nuclear Waste Negotiator, 1993; Easterling & Kunreuther, 1995). In each of these situations, planning grants were given to communities that expressed an interest in hosting the facility without implying a commitment to accept the facility. Rather, the funds were designated to initiate a process so that the community or region would have input into the process and could specify conditions, including compensation arrangements, that would make the site acceptable.

Depending on the type of facility that is being considered, different types of compensation arrangements might be proposed. If a private developer is the applicant, the firm could offer a monetary payment that could be used by the community however it sees fit. Browning Ferris operated in this manner in contacting communities in New York State that might be interested in hosting a landfill through its Community Partnership Program. In 1992, the town of Eagle (with 1,300 residents) overwhelmingly voted in favor of hosting such a facility in return for a benefits package that included tipping fees, local jobs, and free trash disposal worth between $1 million and $2 million (Angell, 1993).

A key aspect of this procedure is that no community is forced to accept a facility against its wishes. This means that it may take a great deal of time to site a particular facility; communities must gain some familiarity and comfort with the concept underlying the facility before they will be willing to enter into negotiations with the developer. In some cases (particularly with respect to radioactive waste disposal facilities), the procedure will not succeed in finding a willing host community. In such a case, it may be necessary for the developer to revisit the choice of technology, examining whether other facilities might be more acceptable to the potential host communities. For the HLNW case, this revisiting of the waste-disposal technology should be performed by a group that includes not only scientists and utility executives, but also representatives of the general public. Opening up this decision process provides the only chance of the selected facility being regarded as legitimate by persons living in a candidate state (Easterling & Kunreuther, 1995).

GENERAL CONCLUSIONS AND RECOMMENDATIONS

In this concluding section we suggest a set of issues that need to be addressed regarding the involvement of the interaction among policy makers, risk management institutions, and the public in dealing with transboundary risk problems facing the public and private sectors. They have relevance to the siting problems facing Hong Kong today that are outlined in the introductory section of the paper.

Higher Quality Public Involvement

Research clearly shows that public involvement is a necessary part of risk management. Research is less clear on the specifics of what that involvement should look like. Though some researchers recommend *greater* public involvement in risk management decisions, we are less certain that more is necessarily better. It is perhaps more appropriate to conclude that high-quality public involvement is more important than, for example, involving more members of the public, or involving the public more deeply in issues that they are poorly prepared to grasp. There is a risk in allowing the public to express their anger and rage but doing very little to accommodate their views or change how things are done. This form of involvement is perhaps better characterized as indulging the public, which sometimes happens under the guise of involving the public.

High-quality public involvement has not yet been well defined. We suggest that risk-management institutions develop guidelines based on definitions of what is wanted by the public and from the public, and how their viewpoints will be incorporated into risk management decisions. Are there technical decisions where public values would be relevant? Can the public be helpful in defining approaches for relating to their own constituency? Is there training and education that the public needs in order to be an active, valued, and respected participant in risk management?

Earlier Involvement of the Public

Very often, the difficulties in dealing with the public are brought about because those impacted by a project are among the last to know of its existence. Project development is a complex and inexact process. For project developers, the road that leads from an idea to a construction permit or operating license is a long and uncertain one. Only a very small number of the projects that are considered actually make it to the point of filing an application with a regulatory or licensing agency.

Usually by the time an application is filed, many decisions have been made that are difficult to reverse, making it difficult, if not impossible, for a proponent to incorporate the public's input. Project proponents need better advice on how to involve the public earlier in the development cycle. And risk management institutions need better guidance on how they can give that advice in a responsible way that is sensitive both to the needs of the public and to the constraints and problems faced by the proponents.

Greater Reliance on "Volunteer Communities"

For the public to be a willing partner in technology, it needs to know what is in it for them. For a project to be of true benefit to a country or a region within the country, it must fit within their own framework of goals and objectives, and not just those of project developers. Project proponents should be encouraged to strive for a partnership with host countries and their neighbors. The first step in establishing that partnership is recognizing the importance of voluntariness in decisions about technology. The "normal" project development process can seem to community members as imposing the results of decisions made by others on them, particularly when public involvement does not occur until far downstream from project planning. By working toward voluntary participation in project development, proponents may actually reduce the risk that a project will run into trouble that can result in costly delays or even more costly abandonment.

Dealing with Affected Communities

One of the major challenges in siting facilities that have transboundary risks is dealing with those communities indirectly affected by the new project. Even nonhazardous facilities, such as landfills, can cause disruptions to neighboring communities by producing odors or increasing truck traffic. Some issues that need to be addressed in this regard are: What are effective ways of dealing with situations where the host community accepts the project but the neighboring communities oppose it? What are the most effective ways to obtain public input regarding the concerns that residents in neighboring communities have with the proposed facility? What role can monetary or in-kind compensation play in diffusing the concerns of these affected communities?

Role of Public Interest Groups

One of the issues that deserves further discussion is the role that environmental groups and organizations can play if a voluntary siting process is used. Suppose that one of these public interest groups feels that it would be inappropriate for community Y to host a proposed incinerator because they feel that the technology is unsafe. The public-interest groups' views of the risk should be presented to the affected residents, who may then revise their feelings about the facility and/or demand additional safety and mitigation measures before agreeing to vote in favor of hosting it in their backyard.

Increase Public Trust

We are currently at an important juncture in the evolution of socially accountable risk management. All of the research to date on the failures of risk management point strongly to the erosion of trust both in government and in many of our social institutions as an important causal factor in the conflicts that exist between the community of risk experts and the public.

We need to move forward in one of two directions. One path advocated by a number of researchers is to work toward increasing public trust in risk management. A second path leads in the direction of developing risk management processes that don't rely on trust, or rely on it only minimally. Though it is seldom acknowledged explicitly, many of the steps currently being taken by government and industry to involve the public through community advisory panels and the like are, in effect, establishing layers of oversight such that the checks-and-balances principles inherent in democratic governments are instituted within technological risk management. This may be a fruitful avenue to pursue, and research along these lines is currently needed.

REFERENCES

Angell, P. (1993). Personal communication regarding Browning-Ferris Industries, Inc.'s program for siting landfills, April (Houston: BFI).

Bacot, H., Bowen, T., & Fitzgerald, M. (1994). Managing the solid waste crisis: Exploring the link between citizen attitudes, policy incentives, and siting landfills. *Policy Studies Journal, 22*(2), 229–244.

Carnes, S. A., Copenhaver, E. D., Sorensen, J. H., Soderstrom, E. J., Reed, J. H., Bjornstad D. J., et al. (1983). Incentives and nuclear waste siting: Prospects and constraints. *Energy Systems and Policy, 7*(4), 324–351.

Dunlap, R. E., & Baxter, R. K. (1988). Public reaction to siting a high-level nuclear waste repository at Hanford: A survey of local area residents. Report prepared by the Social and Economic Sciences Research Center, Washington State University, Pullman, WA for Impact Assessment, Inc.

Easterling, D., & Kunreuther, H. (1995). *The dilemma of siting a high-level nuclear waste repository.* Boston: Kluwer Academic Publishers.

English, M. (1992). *Siting low-level radioactive waste disposal facilities: The public policy dilemma.* New York: Quorum Books.

Ewing, T. F. (1990, July 8). Guarantees near a landfill. *The New York Times.*

Frey, B., Oberholzer-Gee, F., & Eichenberger, R. (1996). The old lady visits your backyard: A tale of morals and markets. *Journal of Political Economy, 104*(6), 1297–1313.

Gregory, R., Kunreuther, H., Easterling, D., & Richards, K. (1991). Incentives policies to site hazardous facilities. *Risk Analysis, 11*(4), 667–675.

Hamilton, J. (1993). Politics and social costs: Estimating the impact of collective action on hazardous waste facilities. *RAND Journal of Economics, 24*(1), 101–125.

Herzik, E. (1993, June 23). Nevada statewide telephone poll survey data. Report presented to Nevada State and Local Government Planning Group, University of Nevada, Reno.

Jenkins-Smith, H., & Kunreuther, H. (2001). Mitigation and benefits measures as policy tools for siting potentially hazardous facilities: Determinants of effectiveness and appropriateness. *Risk Analysis, 21*(2), 371–382.

Jenkins-Smith, H., & Silva, C. (1998). The role of risk perception and technical information in scientific debates over nuclear waste storage. *Reliability Engineering and System Safety, 59*(1), 107–122.

Jenkins-Smith, H., & Stewart, J. (1998). Who will protect my back yard: Dimensions of federalism in political trust. Institute for Public Policy Working Paper Series, University of New Mexico, Albuquerque.

Kasperson, R. (1999, January 7–9). Process and institutional issues in siting facilities. Paper presented at International Workshop on Challenges and Issues in Facility Siting. Academia Sinica, Taibei.

Kraus, N., Malmfors, T., & Slovic, P. (1992). Intuitive toxicology: Expert and lay judgments of chemical risks. *Risk Analysis, 12*(2), 215–232.

Kunreuther, H., & Easterling, D. (1992). Gaining acceptance for noxious facilities with economic incentives. In D. Bromley & K. Segerson (Eds.), *The social response to*

environmental risk. Boston: Kluwer Academic Publishers.
Kunreuther, H., Easterling, D., Desvousges, W., & Slovic, P. (1990). Public attitudes toward siting a high level nuclear waste repository in Nevada. *Risk Analysis, 10*(4), 469–484.
Kunreuther, H., Fitzgerald, K., & Aarts, T. D. (1993). Siting noxious facilities: A test of the Facility Siting Credo, *Risk Analysis, 13*(3), 301–318.
McClelland, G. H., & Schulze, W. D. (1991). The disparity between willingness-to-pay and willingness-to-accept as a framing effect. In D. R. Brown & J. E. K. Smith (Eds.), *Frontiers in Mathematical Psychology*. New York: Springer-Verlag.
McDaniels, T. L., Kamlet, M. S., & Fischer, G. W. (1992). Risk perception and the value of safety. *Risk Analysis, 12*(4), 495–503.
McGlennon, J. (1983). The Alberta experience ... hazardous wastes? Maybe in my backyard. *The Environmental Forum, 2*, 2–25.
O'Connor, R., Bord, R., & Pflugh, K. (1994). The two faces of environmentalism: Environmental protection and development on Cape May. *Coastal Management, 22*(2), 183–194.
Office of Nuclear Waste Negotiator (1993, January). 1992 Annual Report to Congress, Office of the United States Nuclear Waste Negotiator, Boise, ID.
O'Hare, M., Bacow, L., & Sanderson, D. (1983). *Facility siting and public opposition*. New York: Van Nostrand Reinhold.
Rabe, B. (1991). Beyond the NIMBY syndrome in hazardous waste facility siting: The Alberta breakthrough and the prospects for cooperation in Canada and the United States. *Governance, 4*(2), 184–206.
Slovic, P. (1987). Perception of risk. *Science, 236*, 280–285.

4

LULUs, NIMBYs, and Environmental Justice

Bruce Mitchell

INTRODUCTION

The background information for the International Conference on Siting of Locally Unwanted Facilities highlights that "siting locally unwanted land uses (LULUs) is a major policy problem throughout the industrialized world" and that the focus is "to examine the underlying causes for facility siting impasse in Asia and other countries and to suggest ways and strategies to help resolve siting conflicts." In that context, key themes are identified: public participation, consideration of options and alternatives, mistrust and trust building, risk perception, communication and management, incentives and compensation, conflict resolution, approach and strategy, and sharing of experiences—successful and not so successful cases.

The intent here is to (1) examine the relationship between LULUs and NIMBYs and the concept of environmental justice, (2) explore the way in which governments in North America have interpreted and used environmental justice as one means to address issues related to LULUs and NIMBYs, and (3) provide selected examples from Canada to illustrate different approaches to siting LULUs in that country.

LULUs, NIMBYs, AND ENVIRONMENTAL JUSTICE

LULUs AND NIMBYs

Locally Unwanted Land Uses (LULUs) and Not in My BackYard (NIMBYs) are interchangeable concepts used to characterize facilities or services that society collectively requires, but usually does not view as desirable to have in proximity to where people live, work, or play. Examples include landfill sites, incinerators, hazardous waste disposal sites, sewage treatment plants, and airports, all of which are normally viewed as "noxious." Such phenomena have been well recognized and studied for over 30 years in North America (Wolpert, 1976; Popper, 1983; Matheny & Williams, 1985; Ballard & Kuhn, 1996; Lawrence, 1996; Elliott et al., 1997; Elliott, 1998; Ali, 1999; Baxter et al., 1999a, 1999b; Gunderson & Rable, 2000; Rabe, Baker, & Levine, 2000; Wakefield & Elliott, 2000; Hostovsky, 2006; Schively, 2007). Another term sometimes used is TOADS, standing for Temporarily Obsolete Abandoned Derelict Sites (Greenberg et al., 1990, 2000).

In the following subsections, attention turns to two aspects addressed by researchers: principles and procedures.

Principles

In addition to addressing technical matters, investigators have examined principles on which decision makers should base decisions. To illustrate, Baxter et al. (1999a) identified three principles to guide decisions related to LULUs or NIMBYs: trust, equity, and public participation. In their view, *trust* is a key element for relationships among stakeholders, especially government regulatory agencies, siting agencies (public or private), and the host community. It should be added that trust is particularly important when a community volunteers to host a LULU in expectation of economic benefits at the local level. Conflict and opposition often emerge because people in a proposed host community do not trust the regulatory agency or facility proponent (or neither), or the proposed technology. There is a close connection between (lack of) trust and *public participation*: Trust is an end to be achieved, and public participation is one means to achieve that end. Both should contribute to achievement of *equity*, or a fair sharing of risks associated with a LULU.

Public participation is likely to be ineffective when regulatory agencies and siting proponents interpret it to mean only or primarily providing information to the host community about the siting process and possible risks, rather than systematically including the public in the decision making process. In that context, Lawrence (1996) argued that various degrees of participation can lead to different degrees of control by the public, including (1) *procedural control* (influence related to the structure and implementation of the general decision-making process), (2) *locational control* (authority to decide whether to accept a site for a LULU—critical when a community volunteers to be a host), and (3) *facility control* (opportunity to accept the need for, and scale and operating characteristics of, a facility). For many regulatory officials, allocating the above types of control to the host community or general public frequently represents a significant change in power and authority relationships, and some officials can be expected to be unwilling to make such changes. However, a desire or determination by regulatory officials to retain all authority usually is a barrier to building stronger trust and achieving equity. On the other hand, the host community or general public normally is not homogeneous. As a result, when there are basic disagreements between different community groups, the turning over of substantial power to the host community can lead to difficulty in making decisions.

The third principle, *equity*, focuses attention on *fairness* in terms of the social and spatial distribution of environmental risks. A significant challenge, given different needs and interests associated with decisions for NIMBYs or LULUs, is that various kinds of equity are involved. For example, *distributive equity* relates to the distribution of benefits and costs on and among groups in a host community. With regard to costs imposed by a LULU, the normal way to deal with distributive equity is to compensate host communities, usually through financial incentives (tax relief, new or enhanced community facilities). However, by itself, distributive equity usually is not sufficient to meet concerns of host communities. As a result, *procedural equity* initiatives are often used in parallel. These involve modifying processes used for risk prevention, control, and mitigation.

The work by Baxter, Eyles, and Elliott is instructive by highlighting the desirability for decision makers to develop and publicize the basic principles guiding decisions related to LULUs or NIMBYs. Given that such decisions are normally surrounded by emotion and conflict, it is highly desirable, before the decisions are taken, to have agreement on transparent principles upon which analysis, discussion, and decisions will be based. It is almost inevitable that not every stakeholder will get everything wanted or expected related to a decision for a LULU or NIMBY. Nevertheless, experience indicates that if those not getting everything they wanted believe that the process was open and transparent, they were heard, and decisions include action to mitigate or compensate for risk, the likelihood becomes much higher that a site will be found.

Procedures

Attention has been given to creating processes or mechanisms to find sites for noxious materials or facilities. For example, Barbalace (2001) has reviewed a concept proposed by Michael Girrard, a lawyer in New York State, for choosing sites for new hazardous waste facilities. The following discussion is based on her report.

Communities usually oppose or are hesitant about a hazardous waste facility within or adjacent to their community for two main reasons: risk to health and devaluation of property values. On the other hand, a positive view may emerge if communities believe a waste facility will be a trigger to improve both the local economy and quality of life. When the second, positive perspective has dominated, communities sometimes have volunteered to be the host for a waste facility only later to have the state or provincial government refuse permission on the grounds that the state or province would end up receiving a disproportionate proportion of waste generated in the region or nation.

Girrard proposed that a reverse Dutch auction process, as developed by Herbert Inhaler, could be an appropriate tool to help find a site for hazardous waste facilities. Under the reverse Dutch auction process, an auctioneer would propose a minimum amount of compensation for a community that accepted a hazardous waste facility. Barbalace suggested that the minimum bid for such compensation might start at $10 million, and that amount would be advertised for a set period of time (e.g., a month). If no bids were received from a community based on the $10 million compensation, the bid would be raised, say to $20 million, for another set period. Then, if no bid were received, the amount would be raised again, to $30 million, and so on.

Each community would obviously want to receive maximum compensation. If a community were interested but waited too long for the bid to go higher, however, another community might submit a bid and become the host for the site. Hence, there is pressure on potential host communities, as at any auction, to be prepared to make a decision before another community becomes the successful bidder. An important part of the overall process also is for the state to provide a bidding community with sufficient funds to hire an expert to provide advice.

Girrard recognized that not only the community that becomes the host for a LULU could bear negative impacts from a hazardous waste facility. Neighboring communities could be exposed to risk if the hazardous waste were to be transported through them by rail or by truck. The other communities might also be vulnerable from air-borne contaminants from the facility, or to pollution of aquifers or soils if contaminants were to escape from the site and migrate through subsurface processes. Recognizing these possible problems, Girrard proposed a two-step referendum process, following the reverse auction procedure. The first step would involve a referendum for all residents living within a specified radius of the facility site. The second step would be a referendum for all residents of the appropriate local government area in which the facility would be located. If the proposal for a site passed both referenda, a final stage would be an assessment of needs in the community, geological conditions at the site, and other relevant considerations. Such technical considerations and criteria would have to be satisfied before a final decision.

Barbalace commented that the attraction of the reverse Dutch auction process is that it provides a way to ensure a facility is "wanted" by those living in or near its site. As with all procedures, however, the process is not without limitations. In that regard, as Barbalace (2001, p. 2) observed, "The only question would be whether Girrard's plan would achieve environmental justice or entice an impoverished community to accept something that they didn't really want in order to achieve certain economic advantages." A major concern has to be whether a poor community, experiencing closure of key industries and associated high unemployment rates, would offer to be a host for a LULU even when there could be significant risk to health and well-being. The prospects of immediate, short-term gain from employment opportunities or other economic support for the community could override the prospects of risk to long-term health. It is well known that the general population normally finds it difficult to make calculations involving probability, especially for outcomes associated with low-magnitude but cumulative impact.

Barbalace (2001, p. 2) concluded her analysis by stating that "the problem of environmental justice will not be solved overnight." Nevertheless, by examining creative processes to ensure potential sites meet technical criteria and acceptance by residents of a host community and larger region, she hopes that greater sensitization of industries, communities, and societies will lead to ways to identify and establish such sites. In that context, in the next section, attention turns to the concept of environmental justice, in order to explore whether it is a cause of, or possible solution for, LULUs and NIMBYs.

ENVIRONMENTAL JUSTICE

The U.S. Environmental Protection Agency (EPA) defines environmental justice as "the fair treatment and meaningful involvement of all people regardless of race, color, national origin, or income with respect to the development, implementation, and enforcement of environmental laws, regulations and policies."[1] The EPA indicates that "fair treatment" means that no group of people should bear a disproportionate share of negative environmental consequences from industrial, commercial, or municipal operations, or from the implementation of federal state, local, or tribal policies or programs (EPA, 1997). The Office of Environmental Justice in the state of New York uses the

---

1 www.epa.gov

same definition as the EPA, and elaborates that "Environmental justice efforts focus on improving the environment in communities, specifically minority and low-income communities, and addressing disproportionate adverse environmental impacts that may exist in those communities."[2]

In contrast, the Friends of the Earth (FOE) takes a broader perspective, stating that "Environmental justice means: quality of life for all—everyone should have a safe and healthy place to live, work and play; enough for us and the future—we need to make sure there are enough resources for all of us and future generations."[3] The FOE continues, observing that "unfortunately, there are many examples of environmental injustice" and that "it is usually the poorer communities that suffer the most from more pollution … habitat loss … health problems … climate change."

Bullard comments that "The environmental justice movement emerged in response to environmental and social inequities, threats to public health, unequal protection, differential enforcement and disparate treatment received by the poor and people of color. It redefined environmental protection as a basic right."[4] Continuing, he observed that since the mid-1980s environmental justice spread around the world, and "embraces the principle that all communities are entitled to equal protection and enforcement of environmental, health, employment, housing, transportation, and civil rights laws and regulations that have an impact on the quality of life."

It is generally agreed that the concept of environmental justice emerged from a protest in 1982 regarding a hazardous waste landfill site in Warren County, North Carolina, United States which resulted in the arrest of 500 protesters. The protest was against a decision to establish a landfill site for PCB-contaminated soil to be removed from 14 places in the state. The location of the landfill site was adjacent to a small, low-income community whose residents were predominantly African-American. For the protesters, the siting decision highlighted that such hazardous facilities were frequently being located in areas in which the dominant inhabitants were minorities and/or low-income people.

One outcome of the Warren County protest was a study by the U.S. General Accounting Office (1983). It focused on eight southern states, with the goal to determine if there were an association between the location of LULUs or NIMBYs and the racial and economic status of nearby communities. The

---

[2] www.dec.ny.gov/

[3] www.foe.co.uk/resource/faqs/questions/environmental_justice/html

[4] www.ourplanet.com/imgversn/122/bullard.html

General Accounting Office reported that three of every four such landfills were sited in or close to minority communities. Subsequent studies confirmed this pattern, and from them emerged the concepts of environmental justice, equity, racism, and classism.

Later, in 1991, the First National People of Color Environmental Leadership Summit was held in Washington, D.C. This summit attracted over 1,000 delegates from across the world and from 50 U.S. states. On October 27, 1991, the Summit adopted and released 17 Principles of Environmental Justice (Table 1). The following year (1992), these principles were introduced at the Earth Summit in Rio de Janeiro.

During February 1994, President Clinton released Executive Order 12898, entitled "Federal Action to Address Environmental Justice in Minority Populations and Low-Income Populations" (Clinton, 1994). This executive order required that "each Federal agency shall make achieving environmental justice part of its mission by identifying and addressing, as appropriate, disproportionately high and adverse human health or environmental effects of its programs, policies, and activities on minority populations and low-income populations in the United States and its territories and possessions." The executive order did not, however, direct U.S. federal agencies to consider environmental effects of related activities on populations in other countries, an aspect attracting attention as developed countries such as the United States ship noxious wastes for disposal in other countries (Clapp, 1994a, 1994b, 2001). In this way the risk and burdens related to living or working adjacent to LULUs or NIMBYs are shifted to vulnerable populations in other, and usually developing, countries. This aspect is discussed more below.

Responsibility for coordinating interagency initiatives related to environmental justice was given to the Environmental Protection Agency, and in this way the concept was institutionalized within U.S. federal governance structure. Furthermore, Ringquist and Clark (2002, p. 354) noted that at least four states (California [2002], Florida, Texas, Washington) have created environmental justice commissions whose purpose is "to evaluate the degree of environmental inequity in their states and propose changes in environmental policy to reduce any observed inequity." Similar initiatives also have emerged at the local or municipal level. They explained that New York City's "fair share" policy represented best practice, and that its purpose is to ensure that "each borough and neighborhood bears a fair share of undesirable land use burdens and receives a fair share of beneficial public services and amenities" (Ringquist & Clark, 2002, p. 354).

The EPA Office of the Inspector General (EPA, 2004) published an evaluation of the EPA's effectiveness in implementing the intent of the 1994

Executive Order. Its concluded that the EPA "has not identified minority and low-income, nor identified populations addressed in the Executive Order, and has neither defined nor developed criteria for determining disproportionately impacted" ([*disproportionately impacted* is a term used to characterize adverse effects of environmental actions that burden minority and/or low-income populations at a higher rate than the general population] EPA, 2004, p. i). Furthermore, the Office of the Inspector General concluded that the EPA "had not developed a clear vision or a comprehensive strategic plan, and has not established values, goals, expectations, and performance measurements." More positively, however, it was found that the EPA had prepared an environmental justice toolkit, endorsed environmental justice training, and required all regional and program offices to submit "action plans" to create accountability for environmental justice integration.

The Office of the Inspector General (2004, p. ii) also noted that in a response to its draft report, the EPA stated that it "disagreed with the central premise that Executive Order 12898 requires the Agency to identify and address the environmental effects of its programs on minority and low-income populations." Instead, EPA believed the Executive Order "instructs the Agency [EPA] to identify and address the disproportionately high and adverse human health or environmental effects of it [*sic*] programs, policies and activities," and that, in the words of the Office of the Inspector General, "The Agency does not take into account the inclusion of the minority and low-income populations, and indicated it is attempting to provide environmental justice for everyone." (EPA, 2004, p. ii). The report from the Office of the Inspector General is a good reminder that it is one thing to create a law, policy, or regulation. It is quite another to achieve consistent interpretation or implementation related to the stated intent, especially when ideologies shift as a result of new governments taking office. This outcome deserves consideration by any jurisdiction considering development of policies or regulations related to environmental justice to deal with LULUs or NIMBYs.

Environmental justice is not confined to "local issues." As Bryant (2007a, p. 1) has noted, due to increased restrictions on disposal of toxic wastes in developed countries, combined with growing opposition to toxic waste sites, governments and private waste management companies have been seeking alternative sites in other countries. In Bryant's view, the target countries have been "the politically and economically less powerful nations of the world." The attraction for governments in such nations is substantial payments to receive toxic wastes, and opportunity to create employment opportunities in building and operating the waste sites, whether or not the communities volunteered to be hosts for such sites. However, Bryant also observed that governments and peoples in developing countries are increasingly showing less interest in being "dumped on."

Bryant (2007a) has highlighted examples of toxic wastes from developed countries being sent for disposal to developing countries:

- Jamaica refused to accept a shipment of 20,000 bags of milk powder because of unacceptable high levels of radioactivity. The European Economic Community (EEC), the supplier of the milk powder, argued that the bags of milk were safe, and, after Jamaica refused to accept them, the EEC terminated all its food aid to Jamaica. Later, the EEC removed most of the milk powder from Jamaica.
- In a 10-month period, nearly 4,000 tons of toxic wastes were sent to Koto, Nigeria. Subsequently, there was a significant increase in premature births and the incidence of cholera. The leaking barrels holding the toxic wastes were later removed, but they had been buried for over 10 months, resulting in toxic wastes having leaked into the soil.
- The government of Haiti approved an import permit for fertilizer to be dumped into the Khian Sea. However, it was later learned that the "approved" cargo on the freighter included over 13,000 tons of toxic municipal ash from Philadelphia.

Given such examples, and others in Clapp (1994a, 1994b, 2001), advocates of environmental justice have argued that it is not an acceptable solution to deal with national or domestic LULU or NIMBY problems in developed countries by moving the contentious material or facility to a developing country. The examples also emphasize that LULU or NIMBY situations can involve transjurisdictional complications, up to and including relationships between two or more nations.

In the past 10 to 15 years, considerable research has focused on environmental justice (Bryant & Maliai, 1992; Been, 1993; Higgins, 1993; Bullard, 1994, 2000, 2005; Bryant, 1995, 2007b; Cutter et al., 1996; Foreman, 1998; Simon, 2000; Taylor, 2000; Cole & Foster, 2001; Draper & Mitchell, 2001; Mitchell, 2001, 2004; Warner, 2001; Rechtschafren & Gauana, 2002; Lerna, 2005; Vig & Kraft, 2005; Arnold, 2007; Loo, 2007).

Loo (2007, p. 896) has provided a useful summary of this environmental justice research. She observed that research findings indicate advocates of environmental justice normally take a position that environmental problems can only be resolved if basic social, economic, and political inequities are addressed. Pollution and poverty need to be resolved in parallel, if substantive progress is to be realized. Using terms introduced earlier, both distributive and procedural equity require attention.

Loo (2007) also suggests research on environmental justice has evolved. Initial studies considered examples in which there was inequitable distribution of environmental harm and then sought to understand how this happened. Findings and conclusions, however, were not unanimous. Some researchers concluded the inequities were due to intentional discrimination. When that conclusion was reached, attention then turned to whether the underlying cause reflected class or race differences in the receiving communities. Subsequent research provided more sophisticated examination of race and class, with less attention to inequitable outcomes and more consideration to the complex processes that generated environmental injustice. And, in parallel with the shift just mentioned, other researchers started to question the nature and extent of the problem. In Loo's words, "Was environmental injustice really a national or regional problem? Was it really the product of racism or poverty? Or was the siting of environmental hazards simply the outcome of a series of rational economic choices on the part of those involved?" (pp. 897–898).

Against that context of research, in the following section, attention turns to the implications of institutionalizing environmental justice within government in the United States, and what the lessons might be for other countries.

GOVERNMENT APPROACHES TO ENVIRONMENTAL JUSTICE

Ringquist and Clark (2002) have analyzed what they termed the "politics of state environmental justice policy adoption." They provide deep insight into the opportunities and challenges for governments that have decided to use environmental justice to deal with LULUs and NIMBYs. In this section, the observations, findings, insights, and recommendations from Ringquist and Clark are reported.

Ringquist and Clark observed that, in the late 1990s and early 2000s, there was "an explosion of research regarding environmental justice" (2002, p. 351). In their view, however, virtually all of the research emphasized the exposure of poor and minority communities to environmental risks created by siting of noxious landfills. They argued that there had been "neither a comprehensive survey of state environmental justice policy nor a systematic investigation of the forces that prompt state policy activity in this area." Their purpose was to address both of these aspects.

Governments, whether national, state, or local, have shown great variability when responding to environmental justice problems. Part of the reason is due to divisions within the environmental communities related to what is considered by many to be the fundamental nature of environmental justice problems, and, therefore, what are the most appropriate responses.

In that context, Ringquist and Clark observed that "environmentalists appear divided between small, grassroots groups and large, mainstream environmental advocates" (2002, p. 354). As a result, different goals and methods can create a divide between grassroots and mainstream environmentalists. For example, grassroots activists in the United States frequently seek remediation through civil rights litigation, whereas the mainstream environmental organizations normally seek "regulatory stringency" or punitive action against polluters. As Ringquist and Clark have observed, "The tradeoff seems to be efficiency and compensation for the principles of liability and punishment" (2002, p. 355).

Policy Types and Choices

To understand how various groups may perceive, or react to, different policy choices, it is helpful to recognize many categories of policies. Several typologies exist. One divides policies relative to "particular issue areas," such as national security, education, and environment. Another typology differentiates among policies based on who pays for and benefits from them, and on the distribution of benefits and costs. Based on the second approach, domestic policies are characterized as being distributive, redistributive, protective regulatory, or competitive regulatory.

Policies created to facilitate environmental justice are situated at the intersection of different policy types. As Ringquist and Clark (2002, p. 356) explain, environmental justice policies normally concentrate on allowable levels of pollution, or on the location of polluting facilities, and also on the enforcement of "command and control" regulations. In such circumstances, environmental justice policy can be defined as one kind of "environmental policy." And, environmental policy is usually considered to be the ultimate example of a *protective regulatory policy* because its associated programs are created to protect the public by specifying conditions to govern different private actions.

However, environmental justice can also be interpreted as a "social issue." When this view is taken, it becomes a type of *redistributive policy* because such policy and associated programs are designed to manipulate the distribution of wealth, political, or civil rights, or some other aspect valued by social classes or racial groups. Such policy is redistributive because it creates winners and losers; some value is allocated to one group at the expense of one or more other groups.

Most advocates of environmental justice in North America do not seek to redistribute environmental risk from poor and minority populations to wealthy and white people. Instead, their goal normally is to reduce environmental risk to *all* communities. Furthermore, their view is that reduction of pollution is the only viable long-term solution to the inequitable distribution of environmental risk. Nevertheless, given the compelling evidence that more environmental risk associated with LULUs and NIMBYs has fallen on minority populations (defined by race or income), resolving the problem does require some redistribution of the risks. Such redistribution may be of two kinds: (1) absolute, assuming overall pollution is not reduced, and (2) relative, assuming some reduction of pollution.

Given the above, it is possible to analyze the implications of a redistributive policy frequently used to deal with LULUs: resolving environmental injustice by having municipalities or private companies pay compensation to individuals living adjacent to or near facilities that pose an environmental risk to them. Compensation can take various forms, such as one or more cash payments, increased community facilities (better parks or schools), or even relocation of residents. Each kind of compensation involves a redistribution of wealth. In addition, evidence is strong that communities with relatively low levels of political power are more likely to receive LULUs or NIMBYs. As a result, action to alter either the criteria or processes used to find sites for such facilities represents a redistribution of political power from those who gained from the previous system to those who will benefit from the new arrangements. Whatever redistributive arrangements are used, different winners and losers emerge, highlighting the defining feature of redistributive policies.

Implications of Policy Types

The discussion above indicated that environmental justice can be seen as an example of either environmental or social policy. The implications of viewing environmental justice as one or the other are profound, because, as Ringquist and Clark comment,

> How an issue is defined (e.g. whether environmental justice is seen as environmental policy or social justice policy) determines the policy Subsystem within which policy decisions are made. The policy Subsystem in turn helps to determine which participants have access to policy decisions, and the relative power of these participants. Moreover, redefining an existing policy issue can radically alter or even demolish policy subsystems, instigate significant changes in the magnitude and distribution of

> budgetary resources devoted to the policy, and precipitate substantial changes in legislation. (2002, pp. 357–358)

They use the example of nuclear energy policy to illustrate how policy definition and redefinition contribute to determining the politics of policy choice. In their view, atomic energy in the United States was initially viewed to be a component of economic development policy, with a connection to national security policy. The outcome was that nuclear power policy was made "by supportive government personnel in a very closed policy environment" (Ringquist & Clark, 2002, p. 358).

During the 1970s, however, nuclear power was redefined in terms of the kind of policy it represented due to at least four factors: (1) poor performance of nuclear power plants; (2) high cost of electricity produced by the plants; (3) increased appreciation of the risks associated with disposal of nuclear wastes; and (4) the Three Mile Island reactor accident in 1979. As a result of the combined effects of the four factors, "public perception of atomic energy changed from a safe, clean, and cheap source of power to a serious environmental threat that inflicted huge costs on consumers" (Ringquist and Clark, 2002, p. 358). Consequently, in a short time, "Nuclear power policy was redefined as environmental/consumer protection policy, and this redefinition opened the policy subsystem to environmental organizations and groups of anti-nuclear scientists" (ibid., p. 358). In less than five years, the Atomic Energy Commission and the Joint Committee on Atomic Energy were abolished. They were replaced by various competing administrative agencies and congressional committees, and access to the policy-making process became more open and accessible. Subsequently, federal nuclear policy was significantly altered.

The lesson from the above experience is to highlight that, depending on how a policy is defined or framed, various stakeholders have different access to policy makers and to policy-making processes. This insight suggests that when governments consider adopting environmental justice to address LULUs and NIMBYs, there should be careful assessment regarding how environmental justice policy is defined and framed. As Ringquist and Clark concluded,

> Whether environmental justice is defined as a protective regulatory/environmental policy issue or a redistributive/social justice policy issue condition our expectations about the locus of important policy decisions, the subsystem within which policy decisions are made, the relative power of advocacy coalitions within the subsystem, and several other theoretically important factors. (2002, p. 362)

Politics of Policy Adoption Related to Environmental Justice

As attention is given to whether environmental justice might be adopted as policy, Ringquist and Clark (ibid., pp. 363–368) suggest attention should be given to three factors important for policy change at a state level: (1) external political factors, (2) internal political factors, and (3) policy-specific factors. Each is discussed below.

• *External Political Factors*

Decision makers operate in a complicated policy environment, in which external (international or national) events often are influential. For example, the SARS epidemic and 9/11 both have affected policies regarding national security, and have led to changes in protocols for screening at airports and issuing of passports. Another example is the growing awareness of the implications of climate change, and the resulting moves by many governments to be, or appear to be, more "green" in their policies.

With reference to environmental justice, Ringquist and Clark concluded that external political factors are of little significance for policy makers because, in their view, "there have been no significant changes in general 'policy mood.'… nor has there been any significant movement of public opinion with respect to environmental justice" (ibid., p. 365). Their conclusion reflects the domestic situation within the United States. In that regard, it is important to highlight one aspect related to their conclusion of "no significant changes." There has been a growing unwillingness of or hesitation by governments of developing countries to accept noxious waste from developed countries, regardless of payments and other incentives offered.

• *Internal Political Factors*

It is generally accepted by policy analysts that levels of wealth, economic development, and general political ideology are all important in establishing the boundaries for what will be considered by a government. In addition, the characteristics (number of interest groups, diversity of interests represented, relative strength of groups) of the interest group system are important, as interest groups communicate policy expectations and demands to policy makers. Finally, the professionalism and capacity of governance institutions are critical influences.

Internal political factors usually "should be critical determinants of state policy activity in environmental justice" (ibid., p. 366). In particular, state political institutions are important because redistributive and protective regulatory policies usually trigger sharp conflict, requiring decisions to be taken at the highest levels within government. Government institutions that are both professional and capable are most likely to be able to make decisions under such conflict-laden situations. Consequently, highly professional and capable institutions should be able to craft environmental justice policies and implement programs, whether the policy is viewed as protective regulatory or redistributive. At the same time, success is likely to be greater if both the elected government and professional public servants share an ideology supportive of environmental justice. Without such support, the challenge will be much greater. A final contextual aspect can be important. As Ringquist and Clark noted, "Since environmental justice policies will benefit disproportionately members of minority groups, state racial diversity is likely to be especially influential in this policy area" (ibid., p. 367).

• *Policy-Specific Factors*

Several factors have implications for the success of a particular policy. These include the accumulated policy-relevant understanding related to the nature and severity of the problem, and the likelihood of developing effective solutions. States are less likely to address a problem if it is not judged to be severe or if policy makers are unsure how to resolve it. However, accumulated knowledge often gives policy makers confidence that they can deal with a problem.

Ringquist and Clark concluded that, for the United States, there had been no "defining focusing events" related to environmental justice at the state level. As a result, they advise that the key factor for governments contemplating use of environmental justice is the level of accumulated knowledge about related issues, and possible solutions (ibid., p. 367).

Furthermore, they argue that the perceived degree of environmental problem severity will have a major impact on how the policy is conceived and defined (environmental/protective regulatory or social justice/redistributive). In other words, in regard to environmental policy, the severity of an environmental problem is very important. Control regulations are more likely to be introduced in situations of severe air or water pollution, or severe hazardous waste treatment or storage issues. In contrast, if environmental justice is conceived as a redistributive policy, then severity of the environmental problem is less critical because there is little or no association between redistributive policy and problem severity. Indeed, there is considerable evidence to suggest an inverse relationship between need and policy generosity when a policy is viewed as redistributive. That is, a state experiencing more severe problems, such as many citizens in need, often introduces less expansive redistributive policies and programs compared to states with smaller populations with needs.

Significance for Environmental Justice

Ringquist and Clark offer the following conclusions that should have significant implications for government policy makers or private sector initiators of a LULU or NIMBY.

(1) Environmental justice has attributes of both redistributive and protective regulatory policy, indicating that it "has not been defined by policy makers in any convincing fashion."
(2) As a result of the point (1), it is possible that environmental policy has not been defined in a way specific enough to lead to identifiable patterns of policy making. The outcome? "An undefined issue seems likely to evoke mixed and perhaps ineffective policy responses (where it evokes response at all)." Such a situation could help to explain why state environmental justice policies "have not been aggressive, nor have they been especially high in profile. Environmental justice exists in the policy making arena as a shadow rather than as an issue of substance and immediacy."
(3) Another aspect is lack of agreement within the environmental community regarding the nature of basic environmental justice issues, such as what true environmental justice issues and problems are, their extent, and possible remedies. If such differences exist, it is hardly surprising that government policy makers do not hold a unified view about goals and approaches to address environmental justice concerns (ibid., pp. 380–382).

This detailed review of the Ringquist and Clark study (2002) has been provided because it offers deep insights that should be considered by government decision makers. Environmental justice is a possible concept on which a foundation can be constructed to address LULUs and NIMBYs. When doing so, however, policy makers need to be clear what kind of generic policy it will reflect, what contextual conditions exist to improve the likelihood of it being effective, and how to engage with environmental nongovernmental organizations (ENGOs), which will likely be heterogeneous rather than homogeneous and highly variable in their interpretations of environmental justice and their preference for solutions. Without careful attention to such aspects, the likelihood of being able to develop and implement effective environmental justice policies and programs will be low.

CANADIAN APPROACHES TO LULUs AND NIMBYs

As with other nations, in Canada there has been an evolution of approaches to find locations for LULUs or NIMBYs. Maclaren has explained that approaches have moved from "traditional" to what are termed as either "voluntary," "open" or "willing host." In her words, this change has occurred because:

> Siting waste management facilities has become a conflict-ridden process characterized by massive public opposition, disagreement over the environmental impacts of the facilities, and a general lack of faith in the traditional regulatory or "closed" approach to facility siting. Dissatisfaction with traditional methods has led to the emergence of a new approach that emphasises co-operation over conflict. (2004, p. 391)

Maclaren explains there is one key difference between traditional and voluntary approaches. In the traditional approach, a wide-ranging search is conducted across a region with the goal to identify a site that best satisfies technical criteria, without regard to whether the relevant local community has indicated willingness to be the host. In the voluntary approach, in contrast, emphasis is placed on identifying a willing host community that contains at least one site-satisfying technical criteria.

Traditional or voluntary approaches share a common initial stage. A general region or area is identified within which the search for a site is to be conducted. Key factors in determining the extent of the area include where waste is generated, limits on how far the waste can be transported, and whether there are any political boundaries (municipal, provincial, national) across which the waste could not be moved. Once this task is completed, the two approaches diverge in terms of what happens.

Traditional Approach

The second step involves "constraint mapping." Using environmental protection criteria (e.g., hydrological, soils, land use), planners map the entire region to eliminate areas not satisfying minimum thresholds related to the criteria. To illustrate, an area underlain primarily by sandy soils normally would be viewed as unacceptable because sandy soils do not allow for natural containment related to controlling leachates. The areas satisfying the environmental protection criteria become candidates for subsequent investigation.

Step three focuses on detailed analysis of data for those areas that passed the constraint mapping. Another set of screening criteria is used to identify specific potential sites. Examples could be a minimum area of land, and avoidance of areas designated as environmentally sensitive. The fourth step is a comparative assessment of possible sites, with a goal to identify the best site relative to biophysical, economic, and social criteria. Prior to this stage, decisions are taken regarding whether all criteria have the same value or whether they are allocated different weights.

The traditional approach has limitations. First, identification of possible sites and choice of the best site are influenced to a large extent by scientific and technical criteria and considerations, with social and psychological aspects often ignored because of difficulties in measuring them on a quantitative basis. Second, professionals with technical expertise, such as engineers and scientists, normally make most of the decisions throughout the siting process regarding which criteria to use and what their weights will be. This characteristic often alienates the general public, leaving it feeling powerless and not engaged in a meaningful way. A common outcome is that a site that meets all the technical criteria is identified but then is rejected by local communities (often as much due to dissatisfaction over the process rather than due to possible negative environmental risks).

Voluntary, Open, Willing-Host Approach

Sometimes, the voluntary approach is the same as the traditional approach in that area screening is used to narrow the number of possible host communities. The voluntary approach may not use constraint mapping at all, however, or only apply it after a willing host community has emerged. Once this aspect has been determined, the voluntary and traditional approaches are notably different.

The principal feature of the voluntary approach is a deliberate choice to seek co-operation with the general public, as well as to find a site in or adjacent to a willing community. Another fundamental characteristic is that a community can withdraw from the siting process at any stage.

Normally, step one involves regional meetings at which local communities have an opportunity to learn about the proposed facility as well as about the siting process itself. After these regional information meetings, communities have the option of expressing initial interest in being included as a possible host. For those communities expressing interest, more detailed information meetings are arranged. If, after the second round of meetings, elected officials in a community are still interested, then detailed investigations begin to see if there is a suitable site within the community. If no suitable sites are found, a community must drop out of the process. If one or more acceptable sites are found, then community approval must be obtained, such as by a referendum or by public meetings. If a community does give approval, then it becomes a candidate to receive the LULU or NIMBY facility. If more than one community is a possible host, the appropriate level of government with jurisdiction for finding a site decides which site is the best overall.

Conceptually, a process that results in a LULU being sited in a willing or voluntary community is superior to one that results in one being imposed in a community. As with all processes, however, there are weaknesses. As Maclaren highlights, "First, and probably most importantly, there is no guarantee that any community will volunteer to host the facility. If there is no willing host, then the siting process must start again, after considerable time and money have been spent" (ibid., p. 393). A second limitation is that, although the dominant principle is that the process will find a socially acceptable site, it may do so at the expense of not protecting the environment. Third, some residents in the host community will be exposed to or suffer more from the environmental risks than others due to their proximity to it. That is, the facility truly is "in their backyard." Those most negatively affected may be out-voted by the larger number of people who perceive benefits to the community through promised jobs or enhancement of community amenities. In such situations, fairness requires that the needs and concerns of those most at risk be given attention. Fourth, residents of adjacent communities may be concerned that their community will not receive any direct benefits but could be exposed to risks. Fifth, Maclaren identified an "ethical issue." In her words, "Only communities that have the greatest need for the economic benefits of these facilities are likely to consider volunteering. Ultimately, therefore, the poorest communities may be asked to bear the greatest burden for the consequences of activities that take place elsewhere, such as nuclear power generation and industrial production" (ibid., p. 393).

The fifth consideration mentioned above deserves further comment. The residents of the poorest communities may view themselves to be in a desperate economic situation, with limited or no practical options to remedy their situation. Consequently, the poorest communities may be more willing to overlook or ignore the potential risks of hosting a LULU or NIMBY site. With none or few opportunities for economic renewal, it is not far-fetched to imagine individuals deciding the identified risks, presented in terms of probabilities and the distant future, can be set aside in order to attract a facility that will create jobs during construction and operation phases, as well as through supporting services. And, once a poor community has become the host for a LULU or NIMBY facility, such an experience may be pointed out to other communities when the next LULU or NIMBY site is being sought.

Canadian Experience with the Voluntary Approach

Hazardous waste facilities have been successfully sited using this approach in both the provinces of Alberta and Manitoba (Rabe et al., 2000). However, failures also have occurred. Kuhn and Ballard (1998) document the unsuccessful attempt in British Columbia to find a site for a province-wide hazardous waste facility. Based on their analysis, they suggest the voluntary approach failed because of faults in the public consultation process that led to loss of trust in the overall process by residents of two communities that had offered themselves as possible hosts. Ontario also has had some unsuccessful outcomes (Kuhn, 1998; Gunderson & Rabe, 2000; Mitchell, 2004, pp. 564–566; Maclaren, 2004, pp. 393–394). The implication is that while the voluntary approach overcomes limitations of the traditional approach, it also has distributive and procedural weaknesses that provide challenges.

CONCLUSIONS AND IMPLICATIONS

Decisions to locate LULU or NIMBY facilities are usually characterized by conflict and controversy. Ironically, such facilities are needed because of the collective demand generated by societies, yet individuals rarely are keen to have them located adjacent to where they live, work, or recreate.

Research suggests that it is important to identify transparent principles on which siting decisions will be based and to engage local communities from the outset in the decision process. Furthermore, innovative procedures, such as the reverse Dutch auction, offer opportunities to overcome mistrust about regulatory agencies, facility proponents, and technologies.

Environmental justice has emerged due to concerns that LULU and NIMBY facilities too often were being located within or adjacent to minority and/or poor communities. In countries such as the United States, environmental justice has been institutionalized into federal governance arrangements through a Presidential executive order. However, recent evaluations have shown significant divergence in interpretation about what the executive order mandates federal agencies to do, emphasizing the generic challenge of moving from intent to action.

If environmental justice is to be used as a foundation to guide siting decisions for LULUs or NIMBYs, governments must be aware that it can be defined or interpreted as either a protective regulatory or redistributive policy. Each type of policy leads to different values being emphasized and makes the related decision-making processes accessible to a variety of stakeholders. A further complication is the heterogeneity of environmental nongovernmental organizations, leading to different goals and methods being pursued or advocated by various environmental groups.

"Voluntarism" has emerged as one approach to facilitate siting of LULUs or NIMBYs. The defining characteristic of the voluntary community approach is that individual communities decide, after receiving information about the proposed facility, whether they would like to volunteer to be the host. The strength of voluntarism is that social considerations are addressed, and communities become actively engaged in the siting decision from the outset. This characteristic differentiates the volunteer-host approach from the approach emphasizing technical considerations when searching for a site. A major limitation and ethical concern, however, is that the volunteer approach may increase the probability that most LULU or NIMBY facilities will be sited within or adjacent to very poor communities. The main reason for such an outcome is that the poorest communities may view themselves as having few if any economic opportunities, and so are more willing to accept health and other risks in order to attract activity that will create employment within their communities.

Environmental justice is not a "magic wand" or "silver bullet" to resolve the conflict and controversy normally associated with LULUs and NIMBYs. Nevertheless, it is a powerful concept to sensitize regulatory agencies and proponents to the fact that too often such facilities are sited in or beside communities that are marginalized due to lack of wealth, political influence or power, or minority status. The ideas that have emerged from the environmental justice movement, such as the Environmental Justice Principles prepared at the First National People of Color Environmental Leadership Summit in 1991, provide an excellent basis from which to articulate principles to guide siting policy and decisions for LULUs and NIMBYs.

Table 1. Principles of Environmental Justice, from the 1991 First National People of Color Environmental Leadership Summit*

PREAMBLE

WE, THE PEOPLE OF COLOR, gathered together at this multinational People of Color Environmental Leadership Summit, ... do affirm and adopt these Principles of Environmental Justice:

1) **Environmental Justice** affirms the sacredness of Mother Earth, ecological unity and the interdependence of all species, and the right to be free from ecological destruction.
2) **Environmental Justice** demands that public policy be based on mutual respect and justice for all peoples, free from any form of discrimination or bias.
3) **Environmental Justice** mandates the right to ethical, balanced and responsible uses of land and renewable resources in the interest of a sustainable planet for humans and other living things.
4) **Environmental Justice** calls for universal protection from nuclear testing, extraction, production and disposal of toxic/hazardous wastes and poisons and nuclear testing that threaten the fundamental right to clean air, land, water, and food.
5) **Environmental Justice** affirms the fundamental right to political, economic, cultural and environmental self-determination of all peoples.
6) **Environmental Justice** demands the cessation of the production of all toxins, hazardous wastes, and radioactive materials, and that all past and current producers be held strictly accountable to the people for detoxification and the containment at the point of production.
7) **Environmental Justice** demands the right to participate as equal partners at every level of decision-making, including needs assessment, planning, implementation, enforcement and evaluation.
8) **Environmental Justice** affirms the right of all workers to a safe and healthy work environment without being forced to choose between an unsafe livelihood and unemployment. It also affirms the right of those who work at home to be free from environmental hazards.
9) **Environmental Justice** protects the right of victims of environmental injustice to receive full compensation and reparations for damages as well as quality health care.

* Delegates to the First National People of Color Environmental Leadership Summit held on October 24–27, 1991, in Washington, D.C., drafted and adopted seventeen principles of Environmental Justice. Here, thirteen of the principles are provided.

10) **Environmental Justice** considers governmental acts of environmental injustice a violation of international law, the Universal Declaration on Human Rights, and the United Nations Convention on Genocide.
12) **Environmental Justice** affirms the need for urban and rural ecological policies to clean up and rebuild our cities and rural areas in balance with nature, honoring the cultural integrity of all our communities, and provided fair access for all to the full range of resources.
16) **Environmental Justice** calls for the education of present and future generations which emphasizes social and environmental issues, based on our experience and an appreciation of our diverse cultural perspectives.
17) **Environmental Justice** requires that we, as individuals, make personal and consumer choices to consume as little of Mother Earth's resources and to produce as little waste as possible; and make the conscious decision to challenge and reprioritize our lifestyles to insure the health of the natural world for present and future generations.

The Proceedings related to the First National People of Color Environmental Leadership Summit are available from the **United Church of Christ Commission for Racial Justice**, 475 Riverside Dr., Suite 1950, New York, NY 10115.

Source: www.ejnet.org/ej/principles.html, accessed on June 11, 2007.

REFERENCES

Ali, S. H. (1999). The search for a landfill site in a risk society. *Canadian Review of Sociology and Anthropology*, 36, 1–12.

Arnold, C. A. (2007). Planning for environmental justice. *Planning and Environmental Law*, *59*(3), 3.

Ballard, K. R., & Kuhn, R. G. (1996). Developing and testing a facility location model for Canadian nuclear fuel waste. *Risk Analysis*, *16*, 821–832.

Barbalace, R. C. (2001). Environmental justice and the NIMBY principle. *EnvironmentalChemistry.com.* Retrieved April 27, 2007 from www.EnvironmentalChemistry.com/yogi/hazmat/articles/nimby.html

Baxter, J. W., Eyles, J. D., & Elliott, S. J. (1999a). From siting principles to siting practices: A case study of discord among trust, equity and community participation. *Journal of Environmental Planning and Management*, *42*, 501–525.

Baxter, J. W., Eyles, J. D., & Elliott, S. J. (1999b). "Something happened": The relevance of the risk society for describing the siting process for a municipal landfill. *Geografiska Annaler*, *81B*, 91–109.

Been, V. (1993). What's fairness got to do with it? Environmental justice and the siting of Locally Undesirable Land Uses. *Cornell Law Review*, *78*, 1001–1085.

Bryant, R. (Ed). (1995). *Environmental justice: Issues, policies and solutions.* Washington, DC: Island Press.

Bryant, R. (2007a). *International Chronology of Environmental Justice.* Retrieved April 27, 2007 from www-personal.umich.edu/~bbryant/iejtimeline.html

Bryant, R. (2007b). *Environmental advocacy: Working for economic and environmental justice.* Ann Arbor, MI: Bunyan Bryant.

Bryant, R., and Maliai, P. (Eds.). (1992). *Race and the incidence of environmental hazards: A time for discourse.* Boulder, CO: Westview Press.

Bullard, R. D., (Ed.). (1994). *Unequal protection: Environmental justice and communities of color.* San Francisco: Sierra Club Books.

Bullard, R. D. (2000). *Dumping in Dixie: Race, class and environmental quality* (3rd ed.). Boulder, CO: Westview Press.

Bullard, R. D. (2005). *The quest for environmental justice: Human rights and the politics of pollution.* Berkeley, CA: University of California Press.

California Council for Environmental and Economic Balance. (2002). *Environmental justice: Principles and perspectives.* San Francisco: California Council for Environmental and Economic Balance.

Clapp, J. (1994a). Africa, NGOs and the international toxic waste trade. *Journal of Environment and Development*, *3*, 17–46.

Clapp, J. (1994b). The toxic waste trade with less-industrialized countries: Economic linkages and political alliances, *Third World Quarterly*, *15*, 505–518.

Clapp, J. (2001). *Toxic exports: The transfer of hazardous waste from rich to poor countries.* Ithaca, NY: Cornell University Press.

Clinton, W. J. (1994, February 11). *Memorandum on environmental justice*, Public Papers of the President. Washington, DC: Government Printing Office, 241–242.

Cole, L., & Foster, S. (2001). *From the ground up: Environmental racism and the rise of the environmental justice movement.* New York: New York University Press.

Cutter, S. L., Holm, D., & Clark, L. (1996). The role of geographical scale in monitoring environmental justice, *Risk Analysis, 16*, 517–526.

Draper, D., & Mitchell, B. (2001). Environmental justice considerations in Canada. *Canadian Geographer, 45*, 93–98.

Elliott, S. J. (1998). A comparative analysis of public concern over solid waste incinerators. *Canadian Geographer, 41*, 294–307.

Elliot, S. J., Cole, D. C., Krueger, P., Voorberg, N., & Wakefield, S. (1999). The power of perception: health risk attributed to air pollution in an urban industrial neighbourhood. *Risk Analysis, 19*, 621–634.

Elliott, S. J., Taylor, S. M., Walter, S., Stieb, D., Frank, J., & Eyles, J. (1993). Modelling psychological effects of exposure to solid waste facilities. *Social Science and Medicine, 37*, 791–804.

Elliott, S. J., Taylor, S. M., Hampson, C., Dunn, J., Eyles, J., Walter, S et al. (1997). "It's not because you like it any better ...": Residents' reappraisal of a landfill site. *Journal of Environmental Psychology, 17*, 229–241.

Environmental Protection Agency (EPA). (1997). *About environmental justice.* Retrieved from epa.gov/swerops/ej/aboutej/html

EPA, Office of Inspector General. (2004, March 1). *Evaluation Report: EPA needs to consistently implement the intent of the executive order on environmental justice*, Report No. 2004-P-00007. Washington, DC: U.S. EPA.

Foreman, C. H., (Ed.). (1998). *The promise and the peril of environmental justice.* Washington, DC: Brookings Institution Press.

Greenberg, M. R., Lowrie, K., Solitaire, L., & Duncan, L. (2000). Brownfields, toads, and the struggle for neighbourhood redevelopment: A case study of the state of New Jersey. *Urban Affairs Review, 35*, 717–723.

Greenberg, M. R., Popper, F. J., & West, B. M. (1990). The TOADs: A new American urban epidemic. *Urban Affairs Quarterly, 25*, 435–454.

Gunderson, W. C., & Rabe, B. G. (2000). Voluntarism and its limits: Canada's search for radioactive waste-siting candidates. *Canadian Public Administration, 42*, 193–214.

Higgins, R. R. (1993). Race and environmental equity: An overview of the environmental justice issue in the policy process. *Polity, 26*, 291–300.

Hostovsky, C. (2006). The paradox of the rational comprehensive model of planning: Tales from waste management planning in Ontario, Canada. *Journal of Planning Education and Research, 25*, 382–395.

Kuhn, R. G. (1998). Social and political issues in siting a nuclear fuel waste disposal facility in Ontario, Canada. *Canadian Geographer, 42*, 14–28.

Kuhn, R. G., & Ballard, K. (1998). Canadian innovations in siting hazardous waste management facilities, *Environmental Management, 22*, 533–545.

Lawrence, D. (1996). Approaches and methods of siting locally unwanted waste facilities. *Journal of Environmental Planning and Management, 39*, 165–187.

Lerna, S. (2005). *Diamond: A struggle for environmental justice in Louisiana's chemical corridor.* Cambridge, MA: MIT Press.

Loo, T. (2007). Disturbing the peace: Environmental change and the scales of justice on a northern river. *Environmental History, 12,* 895–919.

Maclaren, V. W. (2004). Waste management: integrated approaches. In B. Mitchell (Ed.), *Resource and environmental management in Canada: Addressing conflict and uncertainty* (3rd ed.) (371–397). Toronto: Oxford University Press.

Matheny, A. R., & Williams, B. A. (1985). Knowlegede vs NIMBY: Assessing Florida's strategy for siting hazardous waste disposal facilities. *Policy Studies Journal, 14,* 70–80.

Mitchell, B. (2001). Environmental justice in Canada: Issues, responses, strategies, and actions. *Zeitschrift fur Kanada-Studien, 39,* 24–43.

Mitchell, B. (2002). *Resource and Environmental Management* (2nd ed.). Harlow, UK: Prentice Hall.

Mitchell, B. (2004). Incorporating environmental justice, In B. Mitchell (Ed.), *Resource and environmental management in Canada: Addressing conflict and uncertainty* (3rd ed.) (555–578). Toronto: Oxford University Press.

Popper, F. J. (1983). LP/HC and LULUs: The political uses of risk analysis in land-use planning. *Risk Analysis, 3,* 255–263.

Rabe, B. G., Baker, J., & Levine, R. (2000). Beyond siting: Implementing voluntary hazardous waste siting agreements in Canada. *American Review of Canadian Studies, 30,* 470–496.

Rechtschaffen, C., & Gauna, E. (2002). *Environmental justice: Law, policy and regulations.* Durham, NC: Carolina Academic Press.

Ringquist, E. J., & Clark, D. H. (2002). Issue definition and the politics of state environmental justice policy adoption. *International Journal of Public Administration, 25,* 351–389.

Schively, C. (2007). Understanding the NIMBY and LULU phenomena: Reassessing our knowledge base and informing future research. *Journal of Planning Literature, 21,* 255–266.

Simon, D. R. (2000). Corporate environmental crises and social inequality: New directions for environmental justice research. *American Behavioral Science, 43,* 633–644.

Taylor, D. A. (2000). "The rise of the environmental justice paradigm: Injustice framing and the social construction of environmental discourses. *American Behavioural Science, 43,* 508–580.

U.S. Department of Energy, Office of Environmental Management. (2002). *Environmental justice: Definition.* Retrieved from www.em.doe.gov/public/envjust/definition.html

U.S. General Accounting Office (GAO). (1983). *Siting of hazardous waste landfills and their correlation with racial and economic status of surrounding communities.* Washington, DC: U.S. GAO.

Vig, N., & Kraft, M. (2005). *Environmental policy: New directions for the twenty-first century* (6th ed.). Washington, DC: CQ Press.

Wakefield, S., & Elliott, S. J. (2000). Environmental risk perception and well-being effects of the landfill siting process in two southern Ontario communities. *Social Science and Medicine, 50,* 1139–1154.

Warner, K. (2001). Linking local sustainability initiatives with environmental justice. *Local Environment, 7,* 35–47.

Wolpert, J. (1976). Regressive siting of public facilities. *Natural Resources Journal, 16,* 103–116.

5

Are Casinos NIMBYs?

Euston Quah and Raymond Toh Yude

INTRODUCTION

In March 2004, the government of Singapore announced that it had softened its longstanding opposition to the idea of a casino in Singapore. This proposal sparked a heated debate on the issue of casino development and legalization of casino gambling among Singaporeans.

The controversy of a casino stems from the differences in perspectives: One camp, pragmatic and rational, adopts a cost-and-benefit approach to the issue; another, moralistic and religious, opposes the project on moral grounds as they regard gambling as belonging to the category of vices (alcohol, tobacco, illicit drugs, and prostitution). It is often argued, however, that there are many economic benefits to be reaped—earning tourist dollars, recovering lost underground gambling revenues, and increased tax revenues for the government. There are, however, concerns about the spillover economic costs—lower productivity of workers due to pathological gambling problems, cost of treatment of such addiction, and cannibalizing the retail and hotel sectors outside the casino sector (Hoon & Ho, 2004, December 3). More important, there might be social costs, such as more street crime (burglaries, thefts, and robberies) and white-collar crime (fraud and embezzlement), environmental externalities due to heavier human and vehicular traffic, and an increase in likelihood of dysfunctional families arising from pathological gambling (Tan, T. S., 2004, November 17).

The fact is that Singaporeans are not averse to gambling itself. Currently there are many legal avenues to gamble, such as buying lottery tickets, betting on horses and football, and playing slot machines in club houses. Singaporeans have also been travelling to overseas regional casinos in Genting Highlands (Malaysia), Macau, and Melbourne, and they can choose to gamble by going on cruise ships that go into international waters. Naturally, there is also illegal gambling in the underground gambling dens, which are heavily sanctioned but hard to eradicate. Ronald Tan (2004, September 29) estimated that in 2003 some $5 billion was wagered on legal gambling activities, $2 billion was wagered in underground betting and gambling, and a further $1.8 billion to $2 billion was spent in casinos outside Singapore. Altogether, the amount spent on gambling represents 5.5 percent of Singapore's GDP and is considered to be large for the economy. It seems however that it is not gambling per se that the people are opposed to, but the proximity of a casino to their homes that bothers them. This is a typical Not in My BackYard (NIMBY) attitude; that is, the local communities refuse to accept facilities designed to serve the state's general economic and environmental welfare (McAvoy, 1999).

Noxious facilities are typical examples of NIMBY facilities. Such facilities are not welcomed because they pose a potential health threat, cause pollution, create a nuisance, and impose other economic and social costs on the host communities. The state wants and needs to develop them because they benefit the state as a whole by providing for goods and services that were previously unavailable or they lower the costs of provision for currently available goods and services (Quah & Tan, 2002). Because of this asymmetric distribution of the costs and benefits, it is often extremely difficult to site NIMBY facilities. Consider this, too: casino gaming is an economic good that is in demand across the world. However, many countries still sanction casinos for the same concerns stated earlier, thereby creating a large amount of unsatisfied demand. When a casino is developed in Singapore, it will be supplying a good to the whole world, analogous to a local community hosting a facility that benefits the whole state. However, Singapore will bear the costs and problems associated with the casino, while the world consumers benefit by reaping consumer surplus due to a greater world supply. Therefore, a casino is a NIMBY facility.

It is necessary to qualify that the government is not considering the development of a casino per se but an integrated resort-casino (IR). The proposed IR will be a world-class resort with many facilities that cater to a variety of user groups—retail and dining for those who enjoy shopping and fine dining, entertainment shows for family fun, hotels facilities for tourists, conventional facilities for convention attendees, and most important, the casino for recreational gamblers. The IR seeks to enhance "Singapore's reputation as a premium 'must-visit' destination for leisure and business visitors" (Singapore Ministry of Trade and Industry, 2004) and should not to be regarded as a sleazy gambling den. It is an important investment to the government because the tourism industry contributes to 10 percent of Singapore's GDP and 7 percent of the workforce through direct and indirect channels (Economic Review Committee, 2003). However, industry players found

Singapore is losing its attractiveness as it lacks tourist attractions, a natural environment, cultural attractions, and a vibrant nightlife (Khan & Abeysinghe, 1999). Furthermore, it faces strong competition from the region, such as China, Hong Kong, Thailand, and Malaysia. These countries have increased their investments in the development of infrastructure and tourist attractions to enhance the tourist experience. Therefore, the IR is envisioned to be one of the key tourist products that will enhance the attractiveness of Singapore as a tourist destination and gain access to a previously "unreached" group of tourists.

This chapter does not seek to examine the costs and benefits of the casino, instead we focus on the individual decision-making framework that explains the factors affecting Singaporeans' decision to accept or reject the casino. We address the question of whether the casino is a NIMBY and examine the effectiveness of various conflict resolution mechanisms that could maximize the acceptance rate from an unwilling public

The chapter is organized as follows: the second section discusses the proposal of the casino in Singapore, surveys the literature on NIMBY, and draws further evidence on how a casino might fit the bill of a NIMBY. The third section discusses some of the conflict resolution mechanisms that the government can adopt to reduce the level of opposition to the casino; in the fourth section, we present the results of an empirical study on the questions that we set out to answer; and finally, the fifth section summarizes the results with implications for public policy.

CASINO AND NIMBYs

In this section, we discuss the casino proposal and draw comparisons to the NIMBY concept. The NIMBY concept is widely applied to public projects that are asymmetrically beneficial to the general public and not to the host community. A casino does not fit the bill easily, but in a globalized context, Singapore is supplying a good that is in great demand by the world community that appears underproduced. If Singapore builds a casino, she will have to bear the negative consequence of social problems, such as higher crime rates and lower productivity of labour, while the whole world will stand to benefit.

The IR is envisioned to be an iconic attraction with a comprehensive range of world-class amenities that may include hotels, convention facilities, retail and dining, entertainment shows, themed attractions, and most importantly, a casino. It will be at least in the league of world-class Singapore tourism attractions such as the Singapore Zoological Gardens, Night Safari, and the Esplanade—Theatres on the Bay. This allows the economy to increase its revenue from exporting services, create jobs, stimulate local and foreign investment, and bring about greater economic growth through the multiplier effect. Furthermore, the economic rents that arise from the legalization of a popular activity such as gambling can be captured (Eadington, 1999). Hence, there is much economic benefit to be reaped in the development of an IR with casino facilities.

The NIMBY syndrome describes the specific attitude of the people who do not want certain facilities to be located near their residences. It does not matter whether the facilities are relevant and necessary to the state. There are two main characteristics of a NIMBY (Quah & Tan, 2002). First, there is much involvement of the government in the development process. NIMBY facilities usually receive substantial government subsidies and the government will provide for the conditions necessary for the operation of the facility, such as the acquisition of land for development, legalization of its existence, and the protection of the market by restricting entry of competing firms. The government has to decide on the merits of developing the facility and then decide on the siting location if the facility is expected to be beneficial.

The second characteristic is that the facility has many negative environmental externalities. NIMBY facilities pose harm to the environment where they are sited, such as creating water, air, and noise pollution, destroying the aesthetics of the community, or even causing life-threatening hazards. Examples of NIMBY facilities that have been studied include the both nonhazardous and hazardous types, the former includes schools, hospitals, airports, and landfills; the latter refers to chemical plants, refineries, toxic waste treatment plants, nuclear power stations, and waste disposal. The harm to the environment is usually only borne by the community that hosts such facilities whereas the benefit is reaped by the whole society. This asymmetrical distribution of costs and benefits usually results in the development of a NIMBYistic attitude towards the facility.

Casino as a NIMBY

In the first instance, a casino does not easily fit the NIMBY characteristics described above. However, there is definitely a high level of involvement by the government because it has to legalize casino gambling before the casino can be built. Furthermore, the government has to set up a ministerial committee to assess the proposals and award the contract to develop the IR to the corporation able to conceptualize and deliver a world class IR. A casino also creates negative externalities to a society, including potential destruction of the moral and cultural environment of a clean and honest society. Besides these considerations, the key linkage is that when Singapore develops the casino, she will be the local community host supplying a good to this highly globalized world and she will have to bear the negative externalities while tourists around the world benefit. This is analogous to the typical NIMBY situation where she is the host community that has to bear all the costs while the world on the whole benefits from the facility.

The adverse impacts to society can be summarized as follows: (1) The likelihood of pathological or excessive gambling and its related consequences; (2) Increase in street crime issues linked to casinos, such as burglaries, robberies, prostitution, loan sharking, and drug dealing; (3) Potential of corruption of political bodies and law enforcement bodies; (4) Infiltration of gambling operations by criminals, organized or otherwise (Margolis, 1997). Many of these social ills will impact Singapore, the location where the casino is sited, and not the world in general.

The gambling literature studying the relationship between the social costs of gambling and a casino has been inconclusive. For example, Vina and Bernstein (2002), after studying the relationships among a casino with unemployment, bankruptcy, fraud, and embezzlement, found that the conclusions drawn from data analysis are divided and depended heavily on the methodology applied and the sample of data available. Vina and Bernstein argue that unemployment rates appear to be more closely related to the bankruptcy rate than the introduction of gambling. They observe that many economies had developed casinos to stimulate their deteriorating economies, in other words, bankruptcy was rampant *before* the introduction of casino. Their results also show that a small proportion of individuals with pathological or extreme gambling tendencies possess a higher bankruptcy rate than the national average.

Therefore, the actual social costs of a casino are not clear. However, the development of a NIMBY syndrome is usually not based on the actual cost imposed on the society but on the perceived risks and costs to society (Portney, 1991). A survey on the NIMBY literature reveals that many of the opposed projects are technically safe and that authorities had conducted safety and technical tests to determine the sites that would be most suitable for siting the NIMBY facilities. However, local residents often do not trust the government but they lack the necessary information to make rational judgments, so they perceive the harm to be greater than what the facilities truly possessed. Often, the NIMBYistic attitude arises from an asymmetric information coupled with a general distrust of the government in its decision-making process.

Therefore, if we draw a comparison between a casino and the characteristics of a NIMBY, we can establish certain similarities. Though it is not environmentally hazardous, it is morally repugnant and the asymmetrical distribution of perceived costs and benefits cause the people to develop NIMBYistic attitudes. Frey et al. (1996) also argue that the concept of NIMBY can be applied to issues that affect the community and involve wider moral considerations. A casino can thus be regarded as a NIMBY facility and we can analyze the casino in the same manner as an environmentally noxious facility and consider the issues that it poses such as the need for conflict resolution in siting a NIMBY.

CONFLICT RESOLUTION IN NIMBY SITING

The need to resolve the differences in the points of views of the government and the public is an important area of public policy making. Often conflicts that arise due to the NIMBY syndrome can cause significant delays in the siting process and incur heavy opportunity costs. Furthermore, NIMBY facilities are usually necessary for the development of the state, for example, a waste treatment plant, an incinerator, or a power plant. Without them, the economic and social development of the state could come to a standstill. Therefore, the literature on NIMBY dwells extensively on the conflict resolution strategies and other compromises that can minimize the damage and maximize the acceptance with the least amount of delay in building the proposed NIMBY facility (O'Hare et al., 1983).

Many times it is the decide-announce-defend (DAD) process that causes the most amount of opposition to the siting of a NIMBY. Sometimes the people may develop mistrust for the government, or the local authorities and the authoritarian procedures may cause the public to feel that they are being treated unfairly. These can lead to a higher level of opposition to the project. In a democratic setting, authorities will not proceed with the siting process for fear of political backlash in the withdrawal of political support during elections or the eventual withdrawal of the NIMBY facility from the local community.

There is growing consensus among scholars that a successful siting strategy should include both citizens and experts in the decision-making process. A siting strategy that includes citizen participation is advocated because it ensures the process will be fair and democratic (McAvoy, 1999). Although this will make the decision-making process more complex, it allows the other strategies to work more efficiently, namely, the design of mitigation policies to deal with the perceived risk, benefits, and costs, and the design of compensation packages that may increase the support for the NIMBY project. In environmentally harmful NIMBY facilities, it is also necessary to conduct a proper environment impact assessment with the involvement of the community. However, that by itself is insufficient to get higher acceptance by the local residents as it is the perception of risk that matters. Therefore, public participation is the preamble to successful conflict resolution.

Mitigation Policies

Mitigation measures usually involve some form of redesigning of the facility or improved monitoring and decision procedures. It is meant to reduce the actual and perceived risks of the facility. Whilst public participation is the preamble to conflict resolution, Quah and Tan (2002) argue that public participation can be a form of mitigation policy that governments can adopt and it reduces the amount of asymmetric information leading to natural reductions in perceived risk. The literature also suggests that mitigation is more effective than compensation because it seeks to address the real problems posed by the facility and attempts to clarify and change the unfavourable risk perceptions that may be held by the members of the local community.

In the Singapore casino proposal, the government has attempted to impose social safeguards against the casino gambling component. These social safeguards aim to address problem gambling and "mitigate potential impact on our families, social values and work ethos" (Singapore Ministry of Trade and Industry, 2004). According to the proposed social safeguards, the schemes include, among others, a minimum-age requirement, a membership system for Singapore residents, some self-exclusion programs, certain guidelines on credit extension, a facility to allow setting of voluntary loss limits. These social safeguards would only be imposed on casino gambling and do not apply to the non-gambling components of the IR. It is expected that these safeguards will reduce the possibility of developing some of the social problems mentioned earlier.

One of the most important safeguards is the admission by membership. The proposed membership fees are S$100 for a day membership or S$2,000 annually. This is to ensure that those who patronize the casinos will only be those who can afford to do so. This should allay the fear that people with little discretionary income will gamble away the money that was to meet genuine needs and familial responsibilities (Eadington, 1998). On top of this, there is a proposal for exclusion measures to prohibit those who are likely to develop pathological gambling problem from entering into the casino. These measures by themselves are likely to be insufficient to mitigate the risks, and education on the problems of gambling can only be the long-term strategy. Hence, education is included as a safeguard.

Although the effectiveness of the safeguards is uncertain, the government has also set up a forum for the public to provide feedback on the safeguards proposed and to propose their own. This strategy to solicit feedback from the public and to engage them to take ownership of the problem is likely to increase the support for the casino.

Public Relations

The DAD siting procedure mentioned earlier is one of the quickest ways to develop a political confrontation and ruin public relations. Adopting a DAD procedure shows a lack in sincerity on the government's part to engage the public in policy making. Adopting an announce-discuss-decide (ADD) procedure would be one of the public relations strategies that can increase the level of support. Public input can be properly factored in to the decision-making process and genuine mitigation policies in dealing with the concerns of the people can be put in place. In fact, the decision to build the NIMBY facility can even be scrapped without a loss of political credibility. Therefore, maintaining good public relations is a vital element in the resolution of conflicts.

One advantage of good public relations is that the public can be made aware of the same facts that the state officials know, and the public will see the economic benefits of the facility to the surrounding community as well as the benefit to the state as a whole. Managing media relations can also help provide favorable publicity, thus allowing the government to convince those who oppose the project that they are acting narrowly and subverting the general welfare of the state (McAvoy, 1999). Nonetheless, Young (1990) finds it is difficult for the state officials and technocrats to communicate their knowledge of the risks and hazards to the general public in a constructive and credible manner. He argues that effective risk communication involves keeping the lines of communication open, being honest, consistent, and cooperative on the part of the officials, and engaging journalists to understand how to frame the technical and social dimensions of risk issues. While he concedes that it is impossible to eliminate all the risks that the NIMBY facility poses and it may be impossible for even the most honest and open of managers to change the public's attitude toward industrial risks completely, improved risk communication provides a better chance that the merits of opposing arguments will be heard and sincerely considered.

Monetary Compensation

Conventional economic wisdom emphasizes the role of compensation measures in conflict resolution. It assumes that everything can be assigned a monetary value and compensation can be made so that one is indifferent to the cost incurred. In similar fashion, monetary compensation is meant to reimburse the residents for the damage and harm associated with the NIMBY facility. It is even hoped that monetary compensation can tilt the balance and reduce the amount of opposition. However, monetary compensation can be most effective and decisive only when the harm inflicted by a NIMBY facility is the loss of property values. When it comes to the intangibles that the residents might be attached to, such as the aesthetics of the environment, monetary compensation is less effective (Quah & Tan, 2002).

One reason for the difficulties in designing monetary compensation packages is the problem of valuation. Other than the difficulties in valuing the intangibles, there is also a problem of loss aversion (Kahneman et al., 1991). Theoretically the willingness to pay and the willingness to accept values should not have serious discrepancies, the empirical literature suggests that losses are valued more than gains. This implies that the minimum compensation that people demand to give up a good far outweighs the amount they would be willing to pay to keep or acquire the same good. The result is that the compensation will be very much understated if the economic valuation of the compensation is based on the willingness-to-pay measures rather than the compensation-demanded measures.

There is also a growing literature that argues that monetary compensation is not effective for other moral reasons. O'Hare et al. (1983) find evidence through their case studies that compensation in cash is often viewed as a bribe. Frey (1997) also suggests two key reasons why monetary compensation fails—a bribe effect and a crowd-out effect. The bribe effect reduces the willingness to accept the facility by imposing moral costs on the decision to support the project with compensation. The crowd-out effect reduces the intrinsically motivated support, in other words, individuals who initially supported the NIMBY facility for the benefits to the whole society did so out of the civic duty and public spiritedness; however, when monetary compensation is offered, the intrinsic motivation to support is substituted with a transfer of responsibility to the compensating authorities and the level of support for the project is reduced. In that sense, compensation crowds out public spirit and reduces the acceptance of the facility.

Compensation with Public Goods

Compensation need not be in monetary terms, but can be of goods in kind. Goodwill measures, such as providing better street lighting or building more recreational facilities, can be important to the local residents as it helps to maintain a strong and positive presence within the host community (Kunreuther & Easterling, 1996). Two types of public goods as compensation can be identified, first, a public good that might directly mitigate the specific detriments caused by the public harm, and second, a good that has no association with the harms but will benefit the public in general. Examples of the latter type of public goods include subsidized education and involvement in community development.

Mansfield et al. (2002) argue that public goods may be perceived as a fairer method of compensation and not thought of as bribes as they benefit the society and not the individual. People might feel that the moral responsibility for the negative outcomes of the NIMBY facility be spread throughout the community when the compensation is a public good, thereby easing the burden on any particular individual. It reduces the guilt associated with the compensation and restores the intrinsic motivations of civic duty and public-spiritedness, thereby increasing the chance of acceptance. Various scenarios of public harm related to NIMBYs, such as landfills, airports, and recycling transfer centres, were set up and public goods that were designed to be independent to the public harm were offered as compensation. These test whether there are significant differences between compensation with money and with public goods. In a willing-to-accept framework, they concluded there is evidence that public goods are more valuable than cash in the presence of a public harm. Therefore, public goods or other in-kind compensation is an attractive alternative to monetary compensation for public harm—it is likely to be cheaper than monetary compensation and everyone is made better off. More importantly, it is known to be more effective in overcoming opposition from citizen groups. Therefore, compensation by public good is better than monetary compensation, and the authors observe that this type of compensation had been used in some successful facility sitings.

Conflict resolution allows the government to bridge the gap between its desire to site a NIMBY facility and meeting the expectations of the public. The measures of mitigation, managing public relations, and compensation are often limited in effectiveness when implemented in isolation. It takes a combination of all three strategies to deal effectively with the fears and concerns of the public. As we have argued, suitable mitigation policies depend on the type of NIMBY and must be perceived to be effective before the public will accept the NIMBY. Together with compensation schemes, mitigation policies will play an important complementary role in increasing the local residents' likelihood of accepting a proposed noxious facility. It is important to note that the NIMBY syndrome is a formula for paralysis, not progress, the need to resolve the conflict quickly is essential if society is to develop a rational balance between risk and progress.

THE SURVEY AND SOME EMPIRICAL RESULTS

In this section, we discuss the empirical results of a survey that was done to determine the significant factors in influencing an individual to support the development of a casino in Singapore. We employed a Probit model to determine the significant factors in the individual decision-making framework.

The Survey

The survey was administered to 513 Singapore citizens or Permanent Residents above 21 years of age. Ethnicity was not a major concern in this study but instead, religion plays an important role because of the moralist/religious argument against the casino.[1] The breakdown of the 513 respondents in terms of their religion is given in Table 1, along with the population statistics given by the Singapore Department of Statistics. The sample population is small, and hence the proportion does not correspond exactly with the true population data despite the randomness in surveying. Nonetheless, the Buddhist and Taoist groups remain the biggest population in the sample.

Table 1. Religious Affiliation in the Sample Population

Religion	Sample population		Proportion of population* (%)
	Number of respondents	Proportion (%)	
Buddhism and Taoism	184	35.9	44.2
Christianity (including Catholicism)	126	24.6	18.7
Islam	57	11.1	14.0
Hinduism	34	6.6	4.8
No religion	112	21.8	18.3
Total	**513**	**100**	**100**

* The population proportion figures are calculated based on the breakdown given by the Singapore Department of Statistics (2001). It has been adjusted to account only for those who are above 21 years of age.

[1] Although it has been argued that gambling is entrenched in the Chinese culture, and hence ethnicity seems to be an important factor, the reason for its exclusion is that religious values often overlap with cultural values, especially if the religion is especially prevalent among the ethnicity, for example, Islam among Malays, Buddhism and Taoism among Chinese, and Hinduism among Indians.

The survey was developed according to the arguments for and against the development of a casino and contain key factors that will influence the decision-making process. The survey questions can be found in the Appendix. Two separate surveys were administered—one asked about the development of a casino and the legalization of *casino gambling*,[2] while another asked about the development of an *integrated resort-casino* (IR) and the legalization of *casino gaming*.[3] There are two reasons for doing so. First, it controls for the possible psychological effect in using different terminologies, and second, it tests whether there will be a transformation of good effect. It is possible that when an individual perceives the proposal to be a casino per se, the respondent might be more averse to the idea; however, if the good is viewed as a resort, then the opposition for its development might be reduced. Similarly, gambling is viewed as a vice while gaming is a form of entertainment. Therefore, it is necessary to control for this effect. This effect also has important policy implications and is therefore worth investigating. As for the rest of the survey, the order and phrasing remain the same.

We are interested in the factors that might affect the support for the casino and, other than the respondent's religion, the personal characteristics examined include gender, age, income level, education level, whether one cares about the development of a casino (or integrated resort-casino), whether one has children or plans to have children and whether one gambles, or more importantly, the frequency of participation in legalized gambling. The question that asks about whether one cares about the proposal is meant to control for awareness of the developments—a more-informed person will examine the situation in detail and make an informed choice, while someone who does not care, is likely to decide based on hearsay. The question with regards to having a child (or planning to have one) is to examine the concern for the future, that is, if the respondents had answered "yes," it is possible that they might be more concerned about the future generation and therefore might choose not to support the casino proposal.

---

[2] *Gambling* is defined as "an activity in which a person subjects something of value—usually money—to a risk involving a large amount of chance in hopes of winning something of greater value, which is usually money." (Thompson, 1997, p. 3).

[3] A *game* is defined as a free and voluntary activity, governed by rules, and involves some form of risk in the uncertainty of results (McGowan, 2001). It can be classified by a game of skills or a game of luck. A *sport game* is usually of the former type, while *gambling* in a slot machine is a game of luck. Playing of cards against the dealer is difficult to classify as it involves both elements.

The next set of questions asked for the respondent's perception of the impact of a casino/IR in Singapore, such as whether the casino will lead to more social problems, whether there is a risk of Singaporeans developing a gambling addiction, or if the casino will bring economic growth through tourism and increased employment. A third set of questions seeks to find out the mitigation effects that could have taken place due to: (1) an acceptable level of public participation in the policy-making process; (2) a perception of the effectiveness of the safeguards proposed in controlling the potential social problems; and (3) a perception of the importance of siting location to the respondents. All the variables we are interested in examining are summarized in the list of variables found in Table 2.

The respondents were asked whether they support the proposed project before being asked about the factors that might influence their decision-making process. At the end of the survey, three further questions were asked to determine whether conflict resolution mechanisms can take place. In a willing-to-accept framework, we want to test if the respondents are willing to accept a compensation package in monetary terms or in the form of public goods in return for their support. The compensation in monetary terms is in the form of a tax rebate, calculated as a percentage of their current tax. The provision of public goods are of two types: (1) more social workers who are able to mitigate the risk of social problems and pathological gambling, and (2) pure public goods such as the provision of education, giving of more money to the development of the arts in Singapore, and donations to charitable organizations.

Model Description and Methodology

The set of the survey questions allows us to develop a *latent variable model*, $y_i^* = x_i' \beta + u_i$, where y_i^* is the individual's perceived net benefit of the project after undergoing a mental calculation of the costs and benefits. The matrix x_i contains the personal characteristics and the independent variables that determine the costs and benefits and the mitigating factors of the project, while β is a matrix of estimates that describes the marginal effect of each causal factor. We asked the respondents whether they support the development of the proposed casino or IR, recording the response as a binary variable; *support*, where 1 is for a yes answer and 0 for no. It is reasonable to infer that a respondent would only support the proposed project if the perceived net benefit is positive. With this, we can use the Maximum Likelihood Estimation (MLE) method to estimate the marginal effect of each causal factor and we propose to use a Probit model to do so.

The Probit model specification is given as follows (Wooldridge, 2000, pp. 530–531):

$$\Pr\left(y_i = 1 \mid \boldsymbol{x}_i, \boldsymbol{\beta}\right) = \Phi\left(\boldsymbol{x}_i'\boldsymbol{\beta}\right) = \int_{-\infty}^{x_i'\beta} \left[(2\pi)^{-1/2} e^{-v^2/2}\right] dv$$

Table 2. Summary of List of Variables

Variables	Description	Remarks (if any)
gender	1 if male; 0 if female	
educ	1 if primary and below; 2 if secondary; 3 if diploma; 4 if degree or higher	
income	1 if income is below $2,000; 2 if income is between $2,001–6,000; 3 if income is between $6,001–9,000; 4 if income is more than $9,001	
age	1 if 21–30; 2 if 31–40; 3 if 41–50; 4 if 51–60; 5 if 61 and above	
budtao	1 if Buddhist or Taoist; 0 if otherwise	
christianity	1 if Christian; 0 if otherwise	Effect of religion relative to those without any religion
islam	1 if Muslim; 0 if otherwise	
hinduism	1 if Hindu; 0 if otherwise	
children	1 if has children or plans to have children; 0 if otherwise	Effect of concern for the future
gamble	0 if does not gamble; 1 if gambles once in a quarter or less; 2 if gambles once in a month; 3 if gambles once in two or three weeks; 4 if gambles at least once a week	Expect a positive sign, as a gambler is more likely to support the development of a casino
care	1 if cares about the proposed project; 0 if otherwise	The level of awareness towards the proposed project
addiction	1 if perceives very low risk; 2 if perceives low risk; 3 if perceives moderate risk; 4 if perceives high risk; 5 if perceives very high risk	Perceived risk of Singaporeans developing an addiction to gambling; expect a negative sign, as the higher the perceived risk, the higher the associated social costs

Variables	Description	Remarks (if any)
public_part	1 if perceives enough public participation; 0 if otherwise	Expect a positive sign, as more public participation is likely to bring about more support
growth	1 if expects proposed project to increase tourism, create employment and promote economic growth; 0 if otherwise	Perceived positive economic impact; expect a positive sign
social_prob	1 if very unlikely; 2 if unlikely; 3 if likely; 4 if very likely	Perceived risk of social problems; expect a negative sign, as this represents the perceived costs to the society
safeguards	1 if expects proposed safeguards to work; 0 if otherwise	Perceived effectiveness of safeguard; expect a positive sign
site	1 if prefers Marina Bay; 2 if prefers Sentosa Island; 0 if otherwise	Whether choice of site matters, and the proposed site which is preferred; expect a positive sign as Sentosa Island is farther away from the main island than Marina Bay

We have chosen a Probit model because we assume that the error term in the latent variable model for each individual (u_i), which is independent of all the independent variables, is normally distributed. Since we believe that all the factors important for the debate have been included in the analysis, it is the individual's idiosyncrasy that might lead the respondent to support (or not support) the project. Hence, the assumption of normality is imposed.

With two sets of questions, we use three models: (1) Integrated Resort-Casino (IR) model, (2) Casino model, and (3) Pooled model. In the Pooled model, all the observations are regressed as one equation, and a binary variable, *casino*, acts as an independent variable to allow for differences in the intercepts. All three models have the same independent variables given in the list of variables (Table 2), which also has the expected effect of the relationship of the factors with the support level for the casino, or IR.

Regression Results

Regression results for each model can be found in Table 3 and the standard errors given in parentheses are heteroskedasticity-robust. The first question is to determine whether the multivariate regression function differs across the two groups that were asked the same questions using different terminologies, that is, the IR group and the Casino group. To answer it, a Chow Test[4] was used with the following hypotheses:

H_0: There are no structural differences across the two groups of respondents.
H_1: There are structural differences across the two groups.

The Chow statistic calculated is 1.833[5] and the corresponding *p*-value is 1.75%. This implies that at the 5% level of significance, there is statistical evidence to reject the null hypothesis and conclude that there are structural differences in the two groups of respondents who were asked either to support the development of a casino or an integrated resort-casino.

The rejection of the null hypothesis in the Chow Test means the models that correctly describe the different respondent groups are the IR model and the Casino model and not the Pooled model. Before we discuss the significant factors for each model, the implication of this is that there is a behavioural aspect to the problem and this is important for policy making. It appears that the casino proposal undergoes a psychological transformation simply by changing the name to an integrated resort-casino and the gambling industry into gaming industry. The amount of moral/religious opposition to the casino and gambling can be mitigated simply with a change in perspectives where the respondents no longer view the project as a gambling den but an institution developed to attract tourists and promote growth. Therefore, by supporting it, it is not seen as endorsing the vice but agreeing to the liberalization of the gaming/entertainment industry.

When we compare the significant factors for each model, we find that it tells a consistent picture as above. First, we observe that in the IR model, none of the religious variables (including the *constant*) return with a statistically significant estimate, while only *hinduism* is insignificant under the Casino

---

[4] Chow test statistic is given by $F = \left[\frac{SSR-(SSR_1+SSR_2)}{(SSR_1+SSR_2)}\right]\times\left[\frac{n-2(k+1)}{k}\right]$. (See Wooldridge, 2000, pp. 530–531.)

[5] The values for the three models are $SSR_{IR} = 31.289$, $SSR_C = 24.731$, and $SSR_{Pooled} = 59.680$. The degree of freedom (k) is 17.

model. The intercept has to be interpreted with caution, as it contains not only the component that is fixed among all the respondents, but it also contains the factor of being "nonreligious." For example, the negative sign in the estimates for *christianity* and *islam* in both models means that relative to a nonreligious person, a Christian or a Muslim is less likely to support the project. The non-significance of estimates on the religious factors in the IR model implies that the religious opposition to an integrated resort-casino is weak or even non-existant. While in the Casino model, only *hinduism* is not statistically significant, all other religious factors, including the constant, are statistically significant. When we compare across the models, religious factors will play a more important role in influencing the level of support especially for the Casino model. For a person who feels there is average risk in social problems and addiction to gambling, it is estimated that a Christian or a Muslim is 36 percent to 40 percent[6] more likely to oppose to the development of a casino compared to a nonreligious person; however, in the IR model, a Christian is only 5 percent more likely to oppose the project, while a Muslim is 32 percent more likely to do the same. This implies that *ceteris paribus*, a Christian is more likely to displace his religious objection to the project if the project is an IR.

The important implication of this result is that in the area of public policy, if the government can successfully shift the debate on the proposal from a casino to an integrated resort-casino, then it is more likely to garner support from the public. As suggested by the insignificance of the estimate on the intercept of the IR model, even the nonreligious moralists will be less likely to object to the proposal if the other factors such as the risk of social problems and the effectiveness of the safeguards are taken care of. However, it should be stated that the *p*-value for the coefficient estimate on *islam* in the IR model is 12 percent. Although it is still statistically insignificant, it is the only estimate that exhibits such a small *p*-value among the religious factors; therefore, it is possible that in the true population, Muslims may oppose the project on religious grounds even when the debate has shifted.

The only other statistically significant personal characteristic is the frequency of participation in gambling (*gamble*). This result is expected as

---

6 This is based on the calculation, $\beta^{LPM} \approx 0.4\beta^{Probit}$ where *LPM* refers to a linear probability model. The actual fall in probability depends on the individual and the level taken as reference; however, this calculation allows us to estimate the change in probability when all the other values are taken to be its average. All the calculations in the change in probabilities are calculated as such.

those who like to gamble are more likely to support the presence of a casino. Other factors that are statistically significant for both models are revealed in the variables *addiction*, *social_prob*, and *growth*. These factors are related to the costs and benefits of the project and they are similarly expected to play an important role in influencing the decision-making process. While *social_prob* and *growth* have similar effects on the level of support for both models, it is interesting to highlight that in the Casino model, the fear of a risk of addiction to gambling is strongly negative—an estimated 28 percent drop in the marginal probability on support. It can be implied that the relationship between a casino and the perceived risk of addiction is strongly related while the same relationship for an IR is considerably weaker (a 13 percent fall in probability of support). This is consistent with the earlier conclusion that there is a psychological aspect related to the perception of the casino as good and not the IR.

Table 3. Regression Results for Differing Models

Dependent variable: support			
Independent variables	**IR Probit model (casino = 0)**	**Casino Probit model (casino = 1)**	**Pooled Probit model (pooled data)**
casino			–0.00686 (0.1486)
gender	0.0861 (0.2070)	0.131 (0.2270)	0.0571 (0.1479)
educ	0.127 (0.1386)	0.0599 (0.1497)	0.133 (0.1006)
income	–0.370 * (0.1976)	0.216 (0.2005)	–0.126 (0.1313)
age	0.0491 (0.1076)	–0.0703 (0.1315)	0.0473 (0.08061)
gamble	0.397 *** (0.08943)	0.302 *** (0.08403)	0.297 *** (0.05559)
children	0.239 (0.2468)	–0.397 (0.2603)	–0.0398 (0.1747)
budtao	0.245 (0.2783)	–0.621 ** (0.3045)	–0.238 (0.2022)

Dependent variable: support			
Independent variables	**IR Probit model (casino = 0)**	**Casino Probit model (casino = 1)**	**Pooled Probit model (pooled data)**
christianity	–0.136 (0.3198)	–0.908 *** (0.3346)	–0.584 *** (0.2186)
islam	–0.814 (0.5231)	–0.989 ** (0.4247)	–0.959 *** (0.3174)
hinduism	–0.192 (0.4737)	0.467 (0.5179)	–0.299 (0.3904)
care	–0.190 (0.2212)	–0.506 * (0.2679)	–0.267 (0.1635)
addiction	–0.324 ** (0.1483)	–0.709 *** (0.1659)	–0.446 *** (0.1046)
social_prob	–0.383 ** (0.1625)	–0.455 ** (0.1888)	–0.415 *** (0.1169)
growth	1.116 *** (0.3457)	0.910 *** (0.2855)	0.893 *** (0.2074)
public_part	0.376 (0.2296)	0.562 ** (0.2583)	0.404 ** (0.1671)
safeguards	0.652 ** (0.2233)	0.602 ** (0.2338)	0.666 *** (0.1569)
site	0.248 * (0.1275)	0.135 (0.1349)	0.168 * (0.08751)
constant	–0.0424 (0.8024)	2.474 *** (0.8069)	1.037 * (0.5683)
Number of observations	260	253	513
Percent correctly predicted	83.08	86.17	82.46
Log likelihood	–98.265	–76.337	–187.299
McFadden R–squared	0.451	0.558	0.468

Notes: Standard errors are given in parentheses.

* Test statistic is significant at 10%; ** Test statistic is significant at 5%; *** Test statistic is significant at 1%

The three factors that examine the mitigation effects also proved to be interesting. It is observed that the variable *safeguards* is equally important whether the project is perceived as a casino or an IR. The effectiveness of the safeguards proposed by the government will lead to higher support for the project by 25 percent and it indicates that the government should enforce the proposed rules or even enhance them to increase the level of support, as it is perceived as effective in mitigating the risk of social problems and the development of problem gambling. The variables, *public_part* and *site*, have differing effects for each of the models, as the former is only significant for the Casino model, while the latter is significant only for the IR model. The reason for this observation is possibly that the perceived risk for the casino is higher and therefore the effect of public participation on the support is stronger. With regard to siting the casino, the distance between the casino and local residence is not as important to the people because of the efficiency of the transport system in Singapore and the smallness in size. This is not to say that the distance matters for an IR, but perhaps the reason why site matters is because of the environment—Sentosa Island is definitely more desirable for it has been a well-developed tourist attraction and hence more suitable for the development of an IR. Once again, we can conclude that the mitigation effect is not consistent for each model because each good is perceived differently.

In summary, the costs and benefits factors and some mitigation factors matter, whilst the only personal characteristic that matters is the frequency of gambling. Besides this, moral/religious values apply to the Casino model, implying that there is a difference in the perception of the proposed good. Perhaps one of the reasons why the opposition to the proposed project was so strong and the public debate so heated over this issue was because the media had focused exclusively on the issue of a casino and not an integrated resort-casino. The results show that a casino is viewed as a moral/religious issue, while objectivity and rationality prevail for the integrated resort-casino, even though analytically they are the same. Therefore, it is likely that the NIMBY syndrome arose mainly because of the morality of the project and not from other rational reasons such as the spill-over effects of social costs. This implies that the failure of economic theory to understand moral principles might make it difficult to achieve conflict resolution through negotiation and compensation.

Conflict Resolution and Compensation Mechanism

In the NIMBY literature on conflict resolution, we have seen that support for a project can be garnered not only by mitigating the risk factors but also by a process of compensation. Three types of compensation are offered to survey respondents to determine this. These are (1) monetary compensation through a tax reduction, (2) increasing the number of social workers to deal with the social problems of problem gambling, and (3) increasing in public spending on public projects (e.g., education, development of the arts, giving to charitable organizations), using the tax proceeds from the casino.

Compensation with money or public goods has its advantages, but it might crowd out intrinsically motivated support for such projects (Frey, 1997). Some people might switch their response from support to nonsupport as they suspect that the government might have hidden but negative information and wants to "sweeten" the deal. For example, the public might perceive that the project is more risky than previously expected.

We offered three hypothetical types of compensation packages and asked the respondents again for their support for the development of the project. The descriptive statistics in Table 4 allows us to see the effectiveness of each type of compensation. *Compensation* is a dummy variable which has meaning similar to *support*—it is assigned 1 when the respondent answered "yes" to the support for the project when compensation was offered and assigned 0 for "no." While most respondents maintained their original response, we observed that some respondents changed their level of support when compensation was offered. For example, we observe that the number of positive switches (*support* = 0, *compensation* = 1) for the IR model using the tax rebate is 27. This represents 10.4 percent of the total number of respondents answering the IR survey. However, 21 respondents also switched from support to opposition when the compensation was offered. This means that the net increase in support was merely 2.3 percent. This level of net increase in support apparently does not depend on the type of project proposed as a net increase of 2.4 percent was achieved when the respondents were asked to support the casino's development. Compensation using public goods yielded a higher positive change of 8.1 percent and 8.7 percent, respectively, for the IR model and Casino model, while compensation by increasing the number of trained social workers gave a negative result of 4.6 percent and 5.9 percent, respectively.

To test whether each type of compensation mechanism is likely to lead to a significant change in the rate of support, we conducted a Mean Equality test on the rate of support. The following hypotheses are set up for each type of compensation:

H$_0$: There is no change in the level of support with the offer of compensation, that is, $\overline{compensation_l} = \overline{support}$[7]

H$_1$: There is a change in rate of support with the type of compensation, that is, $\overline{compensation_l} \neq \overline{support}$

Taking into account that there is the high correlation between the two series of observations, that is, there is a high likelihood that a respondent would support the project with compensation if he/she had supported the project without compensation and vice versa, we adjusted the standard errors with the necessary covariance factor[8] and calculate the *p*-value.[9] The *p*-values for the equality test are given at the bottom of Table 4.

Though offering compensation with monetary terms yields a positive change in the level of support, according to the results, the change is not statistically significant. This implies that it is not a useful tool in conflict resolution. There might be several reasons for this observation: First, the strong negative switch is probably caused by a withdrawal of support for the project fearing that the government has hidden information about the social cost of the project; second, the respondents might perceive it as a bribe and thus withdraw their support; third, the income tax rebate may not be a good conflict resolution tool because income tax rates in Singapore are already low; and last but not the least, the public may be sceptical about the sincerity of the government in giving tax rebates as the government is known to be shrewd in balancing its budget and the public may fear that it will take back what it gives through other means.

The net drop in the support from those who had previously supported the proposal is statistically significant for both models when the offer of social workers was proposed. This provides more evidence that crowding out may have taken place, and the respondents withdrew their support for the project suspecting that the project might bring about more social problems than the

---

7 The original distributions of *support* and *compensationt* are binomial. However, the Central Limit Theorem states that when the sample size is large enough, the mean values of *support* ($\overline{support}$) and *compensation* ($\overline{compensation_l}$) can be approximated by the normal distribution.

8 When X and Y are not independent, the standard error is adjusted by the following formula: $$S.E.(\overline{X} \pm \overline{Y})^2 = \frac{S.E.(X)^2}{n_X} \pm \frac{2Cov(X,Y)}{\sqrt{n_X n_Y}} + \frac{S.E.(Y)^2}{n_Y}$$

9 $$p\text{–value} = 2 \times \Phi\left(-\left|\frac{\overline{X} - \overline{Y} - \mu_{H_0}}{S.E.(\overline{X} - \overline{Y})}\right|\right),$$ where μ_{H_0} is the test hypothesis value.

government was willing to admit. Therefore, they felt that it is not worth taking the risk. Hence, this form of compensation is not well received.

All in all, the results suggest that the best form of compensation is to use the tax proceeds to provide for public goods such as education, development for the arts, and giving to charitable organizations. Not only are the positive results largest among the types of compensation mechanisms, they are also

Table 4. Effects of Compensation on Support

Type of compensation	Monetary compensation (income tax rebate)		In-kind compensation (social workers)		In-kind compensation (public projects)	
	IR model	Casino model	IR model	Casino model	IR model	Casino model
support[#] = 1 and *compensation*[+] = 1	97 (37.3%)	89 (35.2%)	88 (33.8%)	82 (32.4%)	108 (41.5%)	101 (39.9%)
support = 1 and *compensation* = 0	21 (8.1%)	19 (7.5%)	30 (11.5%)	26 (10.3%)	10 (3.8%)	7 (2.8%)
support = 0 and *compensation* = 1	27 (10.4%)	25 (9.9%)	18 (6.9%)	11 (4.3%)	31 (11.9%)	29 (11.5%)
support = 0 and *compensation* = 0	115 (44.2%)	120 (47.4%)	124 (47.7%)	134 (53.0%)	111 (42.7%)	116 (45.8%)
support = 1 (%)	45.4	42.7	45.4	42.7	45.4	42.7
compensation = 1 (%)	47.7	45.1	40.8	36.8	53.5	51.4
Net change (%)	**2.3**	**2.4**	**–4.6** *	**–5.9** **	**8.1** ***	**8.7** ***
Mean Equality test (*p*–value, %)	38.7	36.6	8.21	1.27	0.09	0.02

Notes: Percentages of respondents in each category are given in parentheses. The sum of the figures may not add up precisely due to rounding errors.

[#] *support* = 1 if the respondent supports the project without any compensation given, otherwise.

[+] *compensation* = 1 if the respondent supports the project when the compensation is given, 0 otherwise

* Test statistic is significant at 10%

** Test statistic is significant at 5%

*** Test statistic is significant at 1%

statistically significant at 1 percent. With the onset of increasing health care costs, a fund to help subsidize medical costs for the poor and the desolate could be set up with the tax revenue and this might be welcomed by the public. Other innovations may include setting up an education trust fund that gives scholarships to the needy or children of a low-income group, or supplementing the tourism-development fund to provide a steady stream of income

that can be used to maintain and develop tourist attractions of Singapore. This is contrasted with the current practice of ad hoc injections of money into the tourism industry by the government. The provision of public good remains the most effective and preferred method of compensation under the willing-to-accept framework and is consistent with the results discussed in Mansfield et al. (2002).

CONCLUSION

While casinos are not typical examples of a NIMBY facility, they can be shown to have similar characteristics because the social costs are shouldered by the local residents among whom the casino facility is sited, but the larger society benefits. While there is no doubt that there would certainly be some gains from casino gaming, such as increased tourist expenditure, there are also potential social costs, such as lower productivity of regular work labour, the possibility of some erosion of institutional integrity, criminal activities, and problems associated with pathological gambling.

This paper examined the factors affecting the public's decision to support the casino. On behavioural grounds, it is found that with a simple change of the terminology from "casino" to "integrated resort-casino" and "casino gambling" to "casino gaming," the acceptance level increased. The reasons for such a psychological effect need further examination. The survey results also found that moral and religious values affect one's response only in the casino case. The respondents may have regarded gambling as a harmful activity, linked to many types of vices such as drinking, prostitution, and using drugs, but treated gaming as an activity that involves skills and talents and is merely harmless entertainment facilitating economic progress.

Governments might find this result relevant should they decide to pursue projects that draw resistance. By creating a positive image, governments can attempt to steer the public to a neutral, rational ground. In particular, for the development of a casino or IR, when the government is able to focus on the latter, rational factors such as the costs and benefits of the project will prevail and correspondingly, moral objections will be reduced. Whether government can do this successfully depends on how it manages its public relations and how the policy makers present themselves in engaging the public.

Relating to this is the issue of conflict resolution in the siting of NIMBYs. This paper found that mitigation policies can affect the level of support for a casino in Singapore. About 47.3 percent of the respondents answered that the safeguards proposed will be effective. According to the models, on average, the level of support can increase by as much as 25 percent if a person perceives the safeguards to be effective. This suggests that the government could do more to promote the effectiveness of the safeguards by emphasizing them and proposing more measures to increase the level of safeguards that might mitigate the risks associated with the casino gambling. These steps are likely to increase the support for the project.

This chapter further examined the usefulness of monetary compensation. It is clear that monetary compensation is not likely to garner a higher level of support because the main factors in opposing the casino, besides moral reasons, are the fear of social repercussions. Compensating individuals is perceived as a bribe and does not address the issues that concern the public. Compensating with goods or services that mitigate the problems related to the project is possible but not welcomed by the public. The form of compensation most likely to be well received and effective in increasing support is the provision of public goods. Governments can consider proposing certain public goods to be provided using tax proceeds or advise the operators of the casino to provide certain services for the community as a form of goodwill gesture to increase project support.

REFERENCES

Eadington, W. R. (1998). Contributions of casino-style gambling to local economies. *Annals of the American Academy of Political and Social Sciences, 556*, 53–65.

Eadington, W. R. (1999). The economics of casino gambling. *Journal of Economic Perspective, 13*(3), 173–192.

Frey, B. S. (1997). *Not just for the money: An economic theory of personal motivation.* Brookfield, VT: Edward Elgar.

Frey, B. S., Oberholzer-Gee, F., & Eichenberger, R. (1996). The old lady visits your backyard: A tale of morals and markets. *The Journal of Political Economy, 104*(6), 1297–1313.

Hoon, H. T., & Ho, K. W. (2004, December 3). No case for casinos. *Business Times.*

Kahneman, D., Knetsch, J., & Thaler, R. (1991). The endowment effect, loss aversion, and status quo bias. *Journal of Economic Perspectives 5*(1), 193–206.

Khan, H., & Abeysinghe, T. (1999). Tourism in Singapore: Past experience and future outlook. In M. K. Chng, W. T. Hui, A. T. Koh, & B. Rao, (Eds.), *Singapore economy in the 21st century: Issues and strategies.* Singapore: McGraw-Hill.

Koh, A. J. H. (1998). *NIMBY and NIABY facilities: The siting issue.* Unpublished bachelor's thesis. National University of Singapore, Department of Economics, Singapore.

Koh, W. T. H. (2004, November 17). An integrated resort-casino for Singapore: Assessing the economic impact. *Institute of Policy Studies Forum on the Casino Proposal.* Retrieved March 14, 2005 from www.ips.org.sg/events/casino/papers/Winston%20Koh.pdf

Kunreuther, H., & Easterling, D. (1996). The role of compensation in siting hazardous facilities. *Journal of Policy Analysis and Management, 15*(4), 601–622.

Mansfield, C., van Houtven, G. L., & Huber, J. (2002). Compensating for public harms: Why public goods are preferred to money. *Land Economics, 78*(3), 368–389.

Margolis, J. (1997). *Casino and crime: An analysis of the evidence.* Washington, D.C.: American Gambling Association.

McAvoy, G. E. (1999). *Controlling technocracy: Citizen rationality and the NIMBY syndrome.* Washington DC: Georgetown University Press.

O'Hare, M., Bacow, L., & Sanderson, D. (1983). *Facility siting and public opposition.* New York: Van Nostrand.

Portney, K. E. (1991). *Siting hazardous waste treatment facilities: The NIMBY syndrome.* New York: Auburn House.

Quah, E., & Tan, K. C. (2002). *Siting environmentally unwanted facilities: Risks, trade-offs and choices.* Northampton, MA: Edward Elgar.

Singapore Department of Statistics. (2001). *Census of population 2000: Education, language and religion.* Singapore: Department of Statistics.

Singapore Economic Review Committee. (2003). *Report of the Tourism Working Group.* Retrieved March 14, 2005 from www.mti.gov.sg/public/PDF/CMT/ERC_SVS_TSM_MainReport.pdf?sid=130&cid=1293

Singapore Ministry of Trade and Industry. (2004). *Social safeguards for integrated resort with casino gaming.* Retrieved March 14, 2005 from www.mti.gov.sg/public/NWS/frm_NWS_Default.asp?sid=38&cid=2257

Tan, R. (2004, September 29). Be cautious—and have two casinos. *Business Times.*

Tan, T. S. (2004, November 17). Social impact of expanded opportunities for gambling: Comparative figures. *Institute of Policy Studies Forum on the Casino Proposal.* Retrieved March 14, 2005 from www.ips.org.sg/events/casino/papers/Tan%20Thuan%20Seng.pdf

Vina, L. de la, & Bernstein, D. (2002). The impact of gambling on personal bankruptcy rates. *Journal of Socio-Economics, 31*(5), 503–509.

Wooldridge, J. M. (2000). *Introductory econometrics: A modern approach.* Cincinnati: South-Western College.

Young, S. (1990). Combating NIMBY with risk communication. *Public Relations Quarterly, 35*(2), 22–26.

APPENDIX

Survey Questionnaire

1) Are you a Singaporean or Permanent Resident of Singapore?[10] Yes/No

2) Do you care about the legalization of casino gambling[11] and the development of a casino[12] in Singapore? Yes/No

3) Gender: Male/Female

4) Age: 21–30/31–40/41–50/50–60/61–65/above 65

5) Highest education level attained: Primary or below/Secondary/Diploma/Degree or higher

6) Religion: No religion/Buddhism/Christianity/Islam/Taoism/Hinduism/Other

7) Income earned: Below $2,000/$2,001–$6,000/$6,001–$9,000/Above $9,001

8) Dwelling type: HDB (3 rooms or below)/HDB (4/5 rooms, EC)/Private Condominium/Private Property

9a) Marital status: Married/Divorced/Widowed/Single

9b) Do you have children or plan to have children? Yes/No

10a) Do you buy lottery (ToTo, 4D, Singapore Sweep) and/or bet in horse races or football matches? Yes/No

10b) On average, what is the frequency you participate in such activities:
Frequency:____
____

[10] This survey was only administered to those who are Singaporeans or Permanent Residents of Singapore and above 21 years of age.

[11] Another set of surveys was administered with the phrase "casino gambling" replaced by "casino gaming."

[12] Another set of surveys was administered with the word "casino" replaced by "integrated resort-casino."

11) Do you support the development of a casino in Singapore? Yes/No

12) What do you think of the risk that the people of Singapore develop an addiction to casino gambling if there is a casino in Singapore?
Very low/Low/Moderate/High/Very high

13) Do you think that there has been enough public participation over the decision to build a casino? Yes/No

14) Do you expect the development of a casino to increase tourism, to create employment, and promote growth? Yes/No

15) Do you think that a casino in Singapore will lead to more social problems (e.g., higher bankruptcy rate, lower productivity, more broken families)?
Very unlikely/Fairly unlikely/Fairly likely/Very likely

16) Do you think that the safeguards proposed by the government (e.g., restriction of age, strict membership rules, imposition of high entry charges) are likely to work? Yes/No

17a) Do you think that the site for the casino matters? Yes/No

17b) If "Yes" in 17a, which site do you prefer?
Sentosa Island/Marina Bay

18a) Will you support the development of a casino, if the government proposes to increase tax rebate as a result of increased revenue due to the casino? Yes/No

18b) If "Yes" in 18a, what proportion of your current income tax do you wish the rebate to be?
Tax rebate: ______________________________

18c) If "No" in 18a, what is your reason?
Reason: ______________________________
____
____

19) Will you support the development of a casino, if the government increases the number of social workers to deal with the social problems mentioned earlier? Yes/No

20) Will you support the development of a casino, if the government increases its public spending on public projects (e.g., education, development of the arts, giving to charitable organizations), using the tax proceeds from the casino? Yes/No

6

Power to the People! Civil Society and Divisive Facilities

Daniel P. Aldrich[1]

Holding up posters stating "Ban Bioweapons in Boston," residents in Boston's seventh district in the South End have been protesting the National Emerging Infectious Diseases Laboratory being built in their neighbourhood. With three-quarters of its funding from the United States federal government, the construction of the biosafety Level 4 laboratory, which would study the world's most dangerous germs, such as Ebola, has angered local groups and civil rights organizations. Opponents pointed out that the urban, densely populated, primarily African-American and Latino area where the laboratory is located could be at risk from potential leaks and accidents. Proponents of the facility, which has yet to see completion due to eight years of delays, argue that its presence in an urban setting is necessary to ensure timely responses to any future outbreaks of biological agents and because of its proximity to well-trained researchers at Boston University and other local hospitals and research facilities. Residents of South Boston are not alone in their Not In My Back-Yard, or NIMBY, mobilization against controversial or unwanted projects.

Citizens around the world have raised their voices, and sometimes their fists, to protest facilities they envision as bringing negative externalities into their neighbourhoods. Residents resisted attempts to site bullet-train extensions (Groth, 1987), airports (Apter & Sawa, 1984; Feldman & Milch, 1982; Feldman, 1985), nuclear power plants (Nelkin & Pollak, 1981; Lesbirel, 1998;

[1] The author wishes to acknowledge the assistance of Christian Brunelli in the preparation of this chapter. The author carried out the research for this chapter in Japan while on an IIE Fulbright Grant and with financial assistance from the Reischauser Institute at Harvard University.

Aldrich, 2008a), temporary housing post-disaster (Aldrich & Crook 2008), and AIDS hospices (Takahashi, 1998). Anti-project movements opposed attempts to site dams (Hagiwara, 1996), accommodations for asylum seekers (Hubbard, 2005), military bases (Smith, 2000), chemical plants (Broadbent, 1998), jails (Hoyman, 2001; Hoyman & Weinberg, 2006), and even group homes (Clingermayer, 1994) in communities around the world.

Due to the asymmetric distribution of costs and benefits, these projects, even though many are necessary for modern life, engender considerable distrust and hostility (Quah & Tan, 2002). Controversial facilities often create broad, diffuse benefits for society as a whole—power generation, public housing, and waste disposal—but focus their costs on those who live most proximate to the site. Without garbage dumps, electrical power stations, and similar projects, modern life in advanced industrial democracies would be paralysed. While these projects are often seen as essential by proponents, local residents may suffer from psychic disutility because of fears of a leak or accident at a nuclear plant or from noise, water, ground, and air pollution released by manufacturing facilities. More concretely, property values and public health may decline as a result of the facility. The diffuse benefits and focused costs of these projects are the opposite of public goods, such as national defence, which bring with them diffuse benefits and diffuse costs. As a result, some scholars label these projects as (local) "public bads" (Frey et al., 1996, p. 1298n1).

The opponents of proposed facilities may be local citizens driven by a desire to protect their neighbourhood or extralocal individuals who see all such facilities, wherever they may be, as unnecessary and dangerous. Williams and Whitcomb (2007) have underscored how even "Green" politicians can oppose the construction of divisive facilities in their backyards. Massachusetts Senator Edward Kennedy and several other high-profile politicians fought the construction of a wind farm off Cape Cod several miles from their beach homes in Martha's Vineyard. Over time, many scholars have gone beyond seeing land use conflicts as isolated or local phenomena which come about because of misinformed, ignorant, or private-good seeking local residents (see van der Horst, 2007). Scholars have argued persuasively that resistance from citizens can push developers and government decision makers to create better-designed and thought-out policies (McAvoy, 1999). Whether or not we envision resistance to planned facilities as selfish, private-good seeking, or a rational response to unnecessary or dangerous projects, such clashes over public bads have been the target of a great deal of investigation by researchers.

Research on controversial facilities can be divided into three main categories: *particularising approaches*, *universalising approaches*, and *variation-explaining approaches.*[2] Particularizing approaches rest on the assumption that cases of land-use conflict are basically different. Hence, each case is often investigated in isolation from others, and the history and sociocultural environment of the participants are critical in understanding relevant events. Past scholars regularly focused on the institutional settings and psychology of protest groups (Wellock, 1978; Nelkin & Pollak, 1981; McKean, 1981; Touraine, Hegedus, Dubet, & Wieviroka, 1983; Garcia-Gorena, 1999; McAvoy, 1999; Weingart, 2001). These studies create rich, in-depth understanding of individual cases from which broader, testable hypotheses can be generated.

A second approach might be deemed the universalising approach. Scholars adopting this method see all cases of opposition to facility siting as essentially the same; Not in My BackYard conflicts are the result of market failures. Because the "market" for controversial facilities cannot easily or naturally clear itself through the actions of private or public investors and local residents, additional mechanisms are necessary to reach equilibrium. Viewing NIMBY politics as a "dragon to be slain," researchers have proposed economic, experimental, and theoretical solutions such as Dutch auctions (Inhaber, 1998, 2001) or property value guarantees (Smith & Kunreuther, 2001) in an attempt to diffuse or prevent such conflict (Kunreuther & Kleindorfer, 1986; Ehrman, 1990; Brion, 1991; Jenkins-Smith & Bassett, 1994; Rabe, 1994; Quah & Tan, 2002). Other proposals to struggles over Locally Unwanted Land Uses (LULUs) include deliberative democratic mechanisms and citizen referenda (Mitchell & Carson, 1986; Munton, 1996).

A final approach to land use conflict might be categorized as explaining variation. Here, scholars seek to explain the variation in the distribution of unwanted facilities, often using correlation between variables, such as race and the presence of controversial projects (Wolverton, 2002). One branch of this approach, known as environmental racism, links the placement of controversial facilities to discrimination and racism (Falk, 1982; Takahashi, 1998). The strongest form of this genre posits that majority ethnic, religious, and racial groups deliberately place unwanted or controversial facilities in the backyards of minority populations. Rather than proposing theoretical solutions to the issue or engaging in abstract theorizing about the causes for resistance to unwanted projects, these researchers regularly practice "action research" in which they participate in lawsuits and seek to alter public consciousness and business and government practice (United Church of Christ, 1987; Bullard,

---

[2] For an alternate approach to the literature, see Aldrich (2005b) and Aldrich (2008a), chapter 1.

1994, 2000; Cole & Foster, 2001; Abel, 2001). As a result of pressure from environment justice activists, President William Clinton signed Executive Order 12898 on February 11, 1994 to ensure that "[a]ll communities and persons across this Nation should live in a safe and healthful environment."

Another up-and-coming branch within the explaining-variation scholarship has identified social capital as an important factor that can be linked to the patterns of distribution of divisive projects (Hamilton, 1993; Clingermayer, 1994; Aldrich, 2008a). In these studies, the depth of interconnectivity between citizens in a community, more than their race, ethnicity, income, or other relevant characteristics, is most strongly correlated with the presence or absence of nuclear power plants, waste facilities, and similar projects.

CRUCIAL BUT UNDERDEVELOPED ROLE OF METHODOLOGY

Despite a large and growing literature on controversial facilities, there is a sizable gap between the standard theoretical literature on the causes and potential "solutions" to NIMBY responses and a pragmatic, real-world, data-based understanding of where controversial facilities have been placed. That is, we have evidence on a large number of often unsuccessful or infamous cases where private developers or government agencies sought to construct new, controversial facilities (Apter & Sawa, 1984; Broadbent, 1998). But we have fewer studies of broader patterns of exactly where authorities sought to locate facilities, and those analyses that have used large-scale analyses have occasionally been hobbled by poor research designs.

Schively (2007a, p. 263) in her review of literature on controversial facilities has proposed "[m]odifying the types of analysis that are conducted in the LULU-siting review process" so that developers and government officials alike can work with local residents to more smoothly site such projects. Here, I agree with her call for change, but alter its focus from developers and governmental authorities to researchers and scholarship. Social scientists continue to produce large numbers of small-N case studies or articles centred on potential solutions to NIMBY problems. The laundry list of potential fixes for the problems of unwanted facilities includes reverse auctions, guarantees of real estate prices over time, medical assistance and insurance for local residents, and compensation. Despite a plethora of new technologies and methods for gathering and analyzing data on these location-based conflicts, many social scientists have yet to embrace new methodologies and technologies for understanding where developers seek to place controversial facilities.

Another problem with many of the existing studies of controversial facilities is that they have not taken the question "Compared to what?" seriously enough. That is, while some studies claim that certain siting methods were inequitable, for example, arguing that an attempt at constructing a new waste dump was carried out in an undemocratic way, such arguments invoke strong counterfactuals. A counterfactual is a claim that involves situations that were not directly observed because they did not yet occur. For example, if a new waste management facility is placed in the backyard of an identifiable minority, and observers criticize the siting as unfair, researchers would need to understand the other, alternative sites that were possible locations for the project. It could be that all technically feasible sites available to the company or governmental authority were located near the group, in which case the final location for the site reflected the make-up of the technically suitable areas, and not deliberate discrimination. Whatever the specifics of the case, researchers must gather additional information on what could have been. In this way, counterfactual thinking is critical.

Further, as scholars we need to be able to lay out all reasonable explanations for a phenomenon and explain why the argument that we are making is the best. We must be able to provide proof that our theory or explanation for how or why a certain facility was or was not built is better than other, alternative theories—and too many of us are not providing the alternative theories in our discussions of divisive facilities. While it may seem obvious at first glance that an unwanted project was placed in an area because of high levels of poverty at that site, a better test of such a theory of discrimination against the impoverished would pit this argument against theories based on politics, racial and demographic composition, local socioeconomic conditions, technical criteria necessary for the plant, and other potential explanations for site selection (cf. Hamilton, 1993; Aldrich, 2008b). To do so we would need a large body of cases to investigate, either drawing in a full sample of all such cases across the region or nation or a smaller but randomly drawn sample to ensure that our cases are representative of the larger population of phenomena.

The field of political geography offers some tools to ameliorate problems in the field of controversial facilities. Given that the response to proposed controversial facilities almost always varies by geography, with more proximal residents more opposed than those farther away (van der Horst, 2007), and researchers identifying strong relationships between distance from the proposed facility and local attitude (Takahashi, 1998), we need to take geography itself more seriously in our work. Ethington and McDaniel (2007), for example, emphasize the importance of geographically based "strong contextualism" in thorough studies of political agency. More broadly, political geographers seek to tie the behaviours and norms of local residents and potential voters to their neighbours, communities, and spatial networks. They also emphasize that all institutions have a geographically defined character, with "regional, continental, and global footprints that define the character, operation, and resources of those institutions" (ibid., p. 137). Hence racist policies such as Jim Crow laws, political attacks on African-Americans, and challenges to desegregation grew out of specific places and spatially rooted histories in North America. For those scholars involved in research on land use conflict, the broader analytical tools and methodologies of political geographers would be of great benefit.

In understanding issues such as location and demographics, political geographers and social scientists alike have moved to studies based on geographic information systems (GIS) analysis and quantitative analysis of larger numbers of cases. Even without GIS systems data, it is possible to use information on multiple numbers of facility siting cases to extract hard-to-see patterns of policy and correlation. A number of scholars have shown the applicability and usefulness of these techniques in their work on controversial facilities. S. Hayden Lesbirel's (1998) ground-breaking work on nuclear and thermal power plant siting in Japan, for example, used a large-scale dataset from which he was able to investigate the factors that sped up or slowed down the siting processes.

Lesbirel argued that he found "considerable predictability in the nature of participants who become involved in siting disputes, the patterns of those participants' responses to projects, and the relationships of those responses with approval times" (1998, p. 142). By linking the number of months necessary to site and build nuclear power and fossil fuel power plants to factors such as electricity demand, political party composition in the area, and bargaining patterns, Lesbirel used quantitative analysis to reveal broader models of the decision-making heuristics of Japanese utility companies. Importantly, Lesbirel linked the events in Japan to broader theories on negotiation and compensation, and allowed scholars in other field to use and apply his insights.

Other empirically sound analyses have used a combination of geographic, socioeconomic, racial, and political data in their work. Ann Wolverton (2002), for example, used data from the North American Toxic Releases Inventory (TRI) in combination with demographic and industrial variables to investigate claims about connections between location decisions and race. Importantly, her analysis linked detailed racial, income-related, and other socioeconomic data with government-mandated data produced by federal facilities and industry groups, and used that large amount of information to illuminate the connections—and, in some cases, lack thereof—between standard theories on race and locational decision making.

New work in social science has used large datasets based on geographic information to uncover otherwise hidden relationships between place and political mobilization. Cho and Gimpel (2007) investigated the clusters of labour and capital necessary for political campaigns, building on past research that has tied a higher presence of campaign volunteers to metropolitan areas with greater levels of existing networks. Geocoding information on the location of volunteers and contributors to the campaign allowed them to link both historical information—for example, who voted in past elections—with new events, such as the discovery of new contributors and the work of new volunteers. With a geographic dataset linked to political, demographic, and historic variables, Cho and Gimpel were far better equipped to analyse important questions of place than standard, alternative methods.

POTENTIAL EXPLANATIONS FOR LOCATIONAL DECISION MAKING

Drawing from the literature on the siting of unwanted projects, and recognizing the importance of laying out all possible explanations for phenomena, we can identify six common explanations for the patterns by which these types of facilities are sited. Table 1 summarizes the potential reasons why authorities may chose one location over another.

Table 1. Six Explanations for Siting Outcomes

Explanation	Logic	Key siting criteria
Technocratic criteria	Bureaucrats control siting process, overlooking politics and local feedback.	Hydrology, geology, and meteorology
Partisan discrimination	Dominant political party punishes political opponents.	Concentration of political opponents
Racial/ethnic discrimination	Racial/ethnic majority punishes minority.	Concentration of ethnic and racial minorities

Explanation	Logic	Key siting criteria
Economic conditions	Wealthy neighborhoods push away facilities; poorer ones seek potential jobs, taxes, and income.	Socioeconomic status of community
Political intervention	Strong politicians bring home what they see as "pork" or push away "bads."	Number and strength of legislators to intervene in process
Civil society characteristics	Mobilization against facilities depends on quality and capacity of voluntary groups.	Solidarity and relative strength of groups within civil society

Note: Adapted from Aldrich (2008a, p. 27)

Technocratic criteria assume that siting decisions ignore local demographic, political, or economic conditions, and select positions for facilities solely on the objective merits of the location. For example, dams require water-resistant bedrock as a base, nuclear power plants should be adjacent to extant power grids and far from highly populated areas, flat land and strong headwinds are desirable for airports, and so forth. Although developers and political authorities frequently reference technocratic criteria (Jackson & Jackson, 2006), research has shown that once we take such technocratic criteria into account, siting is not random. Developers do make sure that their selected sites pass some minimal threshold of meteorological, geological, and geographic benchmarks, but at that point other factors come into play (Aldrich, 2008a). Put another way, when making decisions, technocratic criteria are satisfied, but not maximized.

Two forms of discrimination theories exist for siting unwanted facilities. The first is political discrimination, and rests on the premise that the political party in power punishes its opponents by placing unwanted facilities in their backyard. In Japan, for example, scholars argue that the hegemonic Liberal Democratic Party (LDP) has deliberately placed nuclear reactors in the backyards of opposition party politicians (Ramseyer & Rosenbluth, 1993).[3] The

[3] There is, at present, no evidence to support this argument, at least in the siting of facilities in Japan (Aldrich, 2008a; Aldrich, 2008b).

second form of discrimination is racial, and argues that racial and ethnic majorities in power use their positions to place unwanted facilities in the backyards of minority groups (Bullard, 1994).[4]

Another theory about the distribution of facilities is that authorities place controversial facilities in underdeveloped and impoverished communities, recognizing that such localities are often more willing to take unwanted projects because of potential income from relevant taxes and employment. Alternatively, wealthier communities can hire better lobbyists, lawyers, and activists to fight off unwanted projects. Empirical results for socioeconomic-based siting have been mixed (Mohai & Bryant, 1992). An alternative approach focuses not on the power of the community, but on the power and influence of the politicians from the relevant area. Politicians, at local, regional, or national levels, can intervene in the siting process if they have sufficient influence. These power brokers may lobby developers and bureaucrats to "bring home the bacon" to their constituency, envisioning projects such as nuclear power plants as benefits, and not negatives (Tanaka, 1972, pp. 102–104). Alternatively, politicians may seek to ensure that their constituents are not saddled with projects they see as "bads," so that Massachusetts Senator Edward Kennedy, for example, opposed the construction of wind farms near Cape Cod (Williams & Whitcomb, 2007).

A final but critical approach to siting rests on the recognition that "social capital, defined by trust and cooperation, facilitates collective action to address community issues" (Poley & Stephenson, 2007, p. 5). The literature on civil society recognizes that social capital and civil society are not distributed evenly across societies or even over urban spaces: certain neighbourhoods hold more densely networked populations, while others are more atomized. Citizens in some areas, such as the Village d'le Est section of New Orleans,

---

[4] Many scholars seeking evidence of discrimination looked at current racial and demographic conditions and assumed that those statistics applied in earlier times, ignoring the actual demographics of the period in which the facilities were sited. When Wolverton (2002) retested the hypothesis put forward by the GAO (Government Accountability Office) and UCC (United Church of Christ), she discovered that "race and the presence of TRI (Toxics Release Inventory) facilities (in Texas) are currently correlated, but [found] no such correlation at the time of siting" (Becker, 2004 p. 4n3). Studies of facility siting continue to use current-day demographic data when studying relationships between ethnicity and siting (cf. Pine et al., 2002). While a disproportionate number of North American facilities are located near minority populations, it is not clear if this is the result of discrimination.

have higher trust in each other and mobilize frequently to work on problems collectively, while in other neighbourhoods citizens may be less likely to work together. Research has long demonstrated that stronger community networks allow them to overcome the typical barriers to collective action (Olson, 1965).

A strong neighborhood or community might be one in which neighbours regularly participate in neighborhood-association meetings, work together to pick up trash in public areas, and meet through social functions such as block parties and local fairs. Another type of strong network might be defined by broad participation in horizontal membership associations, such as fishing and farming cooperatives, parent-teacher associations (PTAs), and sports clubs. Towns, cities, and villages with large numbers of members in such groups engage in regular contact and create relationships with enhanced levels of trust. Localities where these civil society organizations are losing members, either due to aging, depopulation, or a lack of interest, can be seen as targets for developers, who seek acquiescent populations for their facilities. Further, qualitative research has shown that anti-facility organizations with strong internal networks are better able to expand their goals beyond single-issue ones (such as stopping the proposed project) to broader, multi-issue ones (such as environmentalism, civil rights, and so forth) (Shemtov, 2003).

Communities with higher levels of social capital are most likely to strenuously and continuously resist siting attempts. Developers and government officials seeking locations for unwanted facilities seek to avoid delay and controversy and evaluate potential host communities through a variety of mechanisms, including "windshield surveys," local polling, focus groups, conversations with power brokers, and past voting records. Once authorities judge a community's mobilization potential as strong or weak, they will move to place projects in areas envisioned as less likely to successfully resist.

CIVIL SOCIETY

Social scientists have come to take the role of civil society seriously in studies of economic and political development. The work of well-established scholars like Robert Putnam (1993, 1995, 2000, 2007), Jonah Levy (1999), Koichi Hasegawa (2004), and others influenced institutions, such as the World Bank, to promote civil society as a way to fight corruption, improve governmental efficiency, and grow economies. Further, some social scientists have argued that civil society plays a role in processes ranging from decreasing homicides and crime rates (Lee & Bartkowski, 2004) to speeding up recovery from disaster (Nakagawa & Shaw, 2004). Recently, social scientists have begun to use the concept of civil society in studying locational decisions for controversial facilities.

A number of cases have confirmed the importance of social capital in the site selection process. Surveyors use what are colloquially called "windshield surveys" of potential sites for low-level radioactive waste. In such an assessment, the surveyor drives through a community observing signs of poverty and low social capital, such as unmowed yards, trailer parks, and boarded-up windows. In one case, uncovered through court documents, a surveyor had noted the large number of trailers in the area and marked it down as "in" the candidate list (as opposed to being excluded). Here, surveyors paid close attention to local signals of potential resistance to siting plans (Sherman, 2006).

Gabrielle Hecht's work on France (1998, p. 248) uncovered a potential explanation for siting decisions carried out by the French government and the state-owned utility EDF (*Électricité de France*). Her work illuminated regional surveys carried out by the state in the late 1950s that showed that the northwest coast had the most favourable reception for potential nuclear power plants. EDF attempted to build nuclear power plants precisely in the areas identified by earlier surveys as having a more pro-nuclear attitude, such as Britanny. However, despite the survey results, which seemed to indicate that the Normandy area would be less hostile to plans for atomic power, these communities strongly fought against the government's plans. If France, known to be a strong state which regularly "steamrollers" its citizenry, takes local levels of social capital into account, it should be no surprise that authorities in England similarly carried out preselection surveys of potential nuclear power plant host communities when making siting decisions (Rüdig, 1994).

Studies show that firms, wary of additional costs of litigation, regulatory hearings, and damage to their reputation, seek to site new, controversial facilities or expand existing ones in those areas where local communities are more receptive, or at least less hostile to their plans. Using data on the decisions of firms in 156 American counties during the 1980s, Hamilton uncovered a strong relationship between measures of potential resistance in the area and the firm's decision to expand its hazardous waste storage, processing, or disposal facilities. Relying on the voter turnout in presidential elections as a proxy for mobilization potential and, holding constant a number of other factors, such as local capacity surplus, the demand for waste processing, and socioeconomic factors, Hamilton found "that firms processing hazardous waste, when deciding where to expand capacity, do take into account variations in the potential for collective action to raise their costs" (Hamilton, 1993, p. 101).

Other studies have shown that tighter social networks create institutions and policies that can keep out unwanted facilities. Using data on 164

American metropolitan areas, Clingermayer (1994) investigated the density of exclusionary zoning devices, such as prohibitions on the siting of group homes for alcoholics, mentally challenged, and juvenile offenders. Controlling for a number of factors, including home ownership, racial composition, socioeconomic status, and the like, he found that "ward representation variable is positively and significantly related to the exclusion of group homes. This effect is far stronger than that of any other variable in the model" (ibid., p. 987). The factor of ward representation enhances the homogeneity of the community, making it more likely that constituents in that politically defined area are more alike—racially, ethnically, and so on—than unlike. In such areas, research has shown that at least in the short and medium term, social capital is higher and interpersonal networks more dense (Putnam, 2007).

Recent research on post-Katrina New Orleans similarly uncovered strong relationships between social capital and mobilization potential and siting decisions by the city of New Orleans and the Federal Emergency Management Agency (FEMA). Following Hurricane Katrina, local and national emergency officials wanted to begin constructing temporary housing, usually in the form of travel trailers, in and around New Orleans. These trailers were intended for residents' use as they restored their water-damaged dwellings, and also could be used by temporary workers, emergency personnel, and government authorities. However, local residents saw the temporary "FEMA trailers" as an eyesore and a potential hazard to their neighbourhoods and strenuously resisted the siting of such facilities in their backyards.

Against this backdrop, the city of New Orleans and FEMA worked together to draw up lists of potential sites for trailers and trailer parks. Aldrich and Crook (2008) analyzed the data for 114 ZIP codes in and around New Orleans, including proxies for socioeconomic factors, race, floodwater damage, education, and home ownership. Only one variable had a strong, significant relationship with the number of trailers slated to be placed in the ZIP code: voter turnout in the presidential elections. Local communities with stronger voluntarism, as demonstrated by their propensity to turn out to vote, were exactly the communities that received plans for fewer trailers. Aldrich and Crook argue, as Hamilton (1993), Putnam (2005), and others do, that higher turnout for elections reflects altruism and voluntaristic tendencies, which are signs of denser social networks and deeper social capital.[5]

[5] Voting, even in advanced industrial democracies, requires the expenditure of time and brings with it opportunity costs both professionally and personally. The actual benefits of voting are hard to quantify, and for many citizens, hard to justify. Hence scholars understand that those who turn out to vote are demonstrating an "other-focusedness" by their willingness to expend their resources on a small return.

Those areas in New Orleans that were less densely populated by volunteers and altruistic types were the areas selected by the city to receive the brunt of the trailer burden, as shown below in Figure 1. The solid line represents the predicted number of trailers, per ZIP code, based on voter turnout, holding other factors, such as race, socioeconomic conditions, and flood damage constant. The dotted lines represent the 95 percent confidence intervals around the estimate. A ZIP code where the majority of the population—80 percent—turned out to vote was slated to receive only 100 or so trailers, while an area that had very few—30 percent or fewer—would receive ten times as many mobile homes. It is not just North American authorities that take civil society into account when siting controversial facilities, however.

In Japan, governmental authorities have been strong and consistent entrepreneurs in the field of commercial nuclear power. Although private utilities handle the majority of responsibilities for planning, constructing, and managing Japan's 55 nuclear power plants, the state subsidizes the industry with favourable loans and research grants, amortizes risk through a variety of insurance and re-insurance mechanisms, and channels hundreds of millions

Figure 1. Trailer Placement Linked to Strength of Civil Society

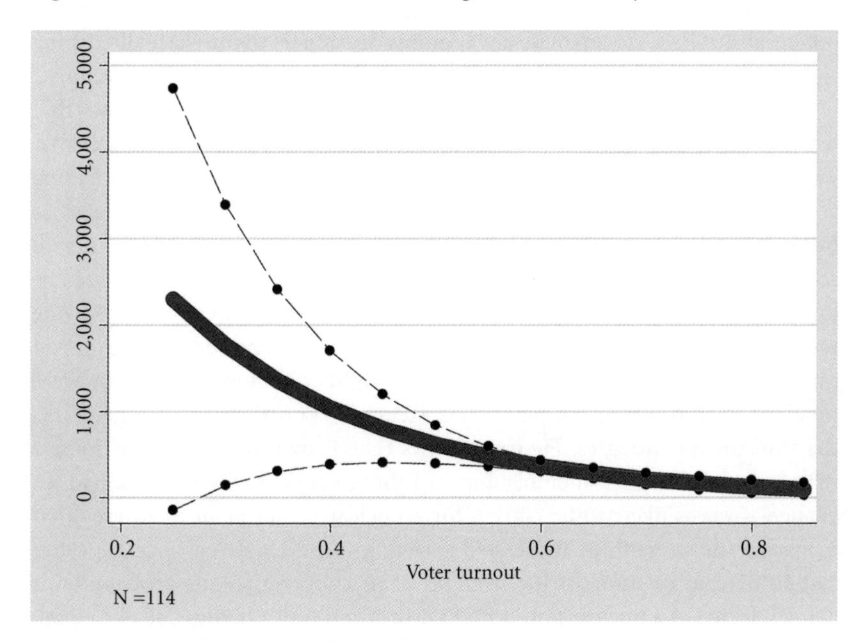

Note: Data from Aldrich & Crook (2008)

of dollars in funds from a submerged tax on electricity use to funds that are distributed to host communities for nuclear power plants. Furthermore, the national government assisted the utilities by surveying potential host communities around the nation. The state has created these various policy instruments in the face of widespread fear of and opposition to nuclear power, an understandable response given Japan's three fatal experiences involving nuclear weapons: Hiroshima, Nagasaki, and the "Lucky Dragon" fishing boat incident in 1955. Against a broad "nuclear allergy" (*kaku arerugi*), the state has supported private electric power companies (EPCOs) in their search for suitable sites for atomic reactors.

Beginning in the late 1950s, and continuing through the 1970s, the Ministry of International Trade and Industry (*Tsūshōsangyōshō*, then known as MITI, in 2001 its name was changed to the Ministry of Economy, Trade, and Industry, or METI) helped survey potential host communities. Internal documents revealed that planners looked closely at the strength of local civil society organizations, especially fishing and agricultural cooperatives (known in Japanese as *nōgyō* and *gyogyō rōdō kumiai*). Decision makers noted closely the density of these associations near potential host communities, recognizing that more powerful, better-connected groups would likely be the first to organize against planned nuclear power plants in their vicinity. Furthermore, planners were concerned about those fishing cooperatives located closest to the potential site, because Japanese law requires developers to purchase the fishing rights to the area of the ocean from which plants would draw cooling water.

A quantitative analysis of close to 200 localities across Japan shows that when utilities and government authorities encountered areas with weak or weakening civil society organizations, as seen by the strength of local fishing cooperatives, they were far more likely to select such towns and villages as hosts for nuclear power plants. In this statistical analysis, the data on local civil society organizations came from the fishing cooperative most proximal to the host village. While it is true that transboundary risk issues could provoke responses from communities farther from the nuclear power plant, in Japan only those cooperatives that hold fishing rights to the nearby ocean were required to sign off on the plant deal. Hence it is on those fishing networks that this analysis focuses. Figure 2 shows that a town or village that merely maintained its original membership in fishing cooperatives—that is, had no change—was as likely to be chosen for a nuclear power plant as an area that increased the strength of those civil society groups. However, once a locality lost 30 percent or more of the strength of its cooperatives, its probability of being selected as a nuclear power plant host community, holding all else equal, increased one hundred-fold.

It was precisely these fishing cooperatives and other local civil society groups that, when intact and powerful, could stall the siting of nuclear power plants for years (Lesbirel, 1998). For example, Iwaishima fishing cooperatives in the village of Kaminoseki helped lead the resistance against planned nuclear power plants there for decades. As in Figure 1, the solid line at the centre of Figure 2 is the predicted probability, and in this case, is the potential that any technically suitable locality has for being selected as the host for a nuclear power plant. The dotted lines on either side of the solid line are the 95 percent confidence interval, as explained.

Japanese government planners used a similar set of decision-making tools when planning sites for airports. As the government discovered at the Narita Airport case, local residents were often upset at the noise pollution generated by the sound of enormous jet engines over their heads (Apter & Sawa, 1984), and this resistance has been seen in France (Feldman & Milch, 1982; Feldman, 1985) and around the world. In fact, to avoid conflicts with local civil society, planners in Japan have sited all of the most recent airports—including Kansai and Chubu airports—offshore, while developers in the United States placed

Figure 2. Decreasing Civil Society Makes Localities More Attractive to Developers

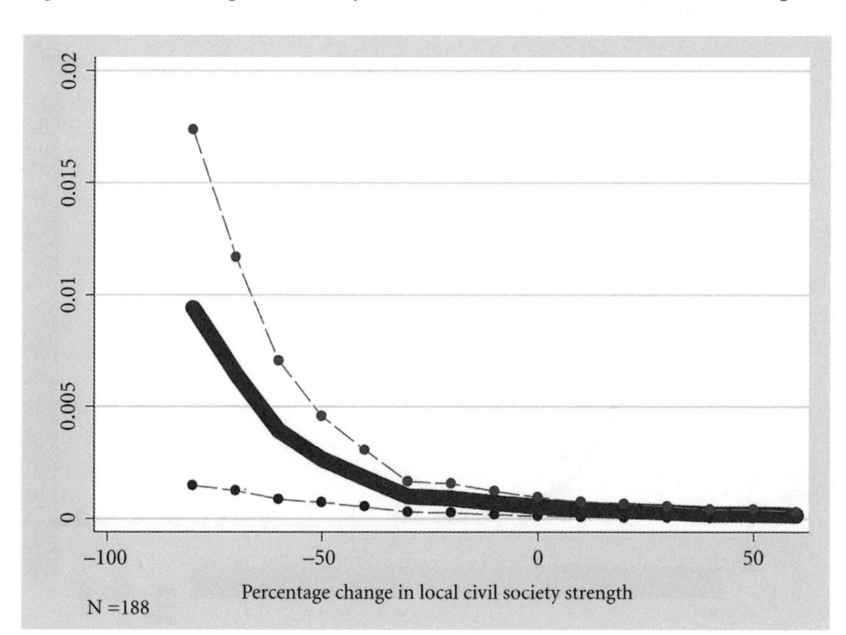

Note: Data from Aldrich, 2008a and Aldrich, 2008b

new airfields far from populated areas that could stymie the process through protest (Altshuler & Luberoff, 2003, p. 162).

As Figure 3 demonstrates, a locality in Japan that simply maintained or even increased the strength of its local civil society organizations—local fishing and farming cooperatives in these cases—was highly unlikely to be chosen as the location for a large-scale, government-managed airport. However, when such a locality began to lose members in those civil society organizations, the state considered it a far better target. Hence the probability that a town or village that has lost more than 50 percent of its farmers and fishermen has for being selected as a host community for an airport is around 1 percent, while a locality which has lost upwards of 80 percent has a 20 percent chance of being chosen.

Given the plethora of evidence that authorities, both governmental and private, take civil society seriously when choosing sites for controversial facilities, the next question is obvious: are these decision makers choosing wisely? It could be that, while they imagine that an area rich in interpersonal connections will be a poor target for an unwanted project, it might not resist any more than an area with sparse social capital. Using 208 attempts to build

Figure 3. Weaker Civil Societies More Likely to Be Chosen as Hosts

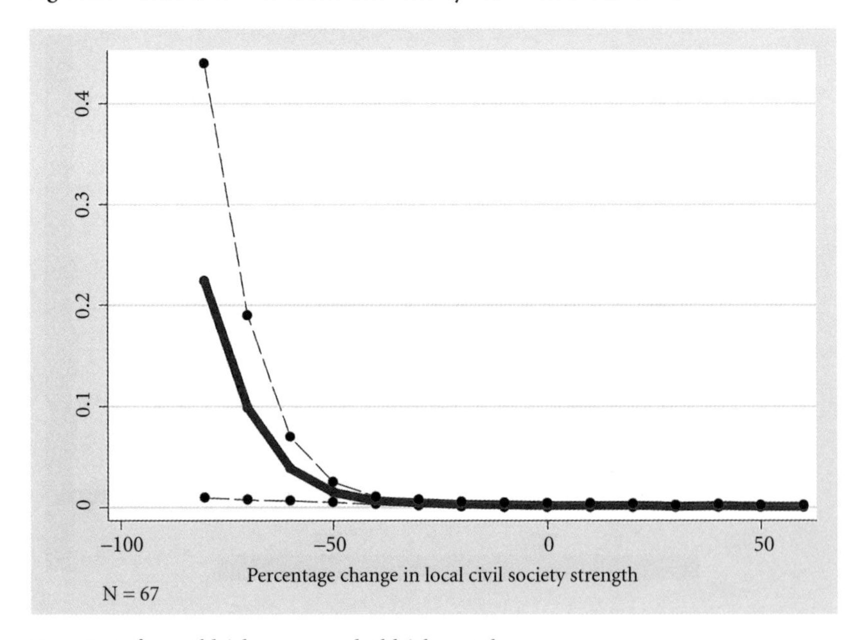

Note: Data from Aldrich, 2008a and Aldrich, 2008b

nuclear power plants, dams, and airports in the latter half of the twentieth century, I analyzed the outcome based on the strength of local civil society group membership. As Figure 4 displays, an area that was able to maintain or increase the strength of its membership in fishing and farming cooperatives was far better positioned to fight off attempts at controversial facility placement. As in the previous figures, the strong solid line at the centre is the predicted probability of the quantity of interest, namely the chance that the proposed facility will be completed.

A town or village that had lost essentially all its membership in civil society associations—meaning a loss in the potential of that area to mobilize and defend itself from proposals for public bads—faced a daunting prospect. Facilities in these localities had close to a 90-percent chance of coming to completion. Completion here is defined as the event where, for example, a nuclear power plant would come online and produce power, a dam would begin to generate hydroelectricity, or an airport would open to planes. On the other hand, where social movements and associations increased in strength over time, the probability of a proposed facility being completed dropped tremendously, to only 20 percent.

Figure 4. Strength of Civil Society Predicts Success or Failure of Public Bads

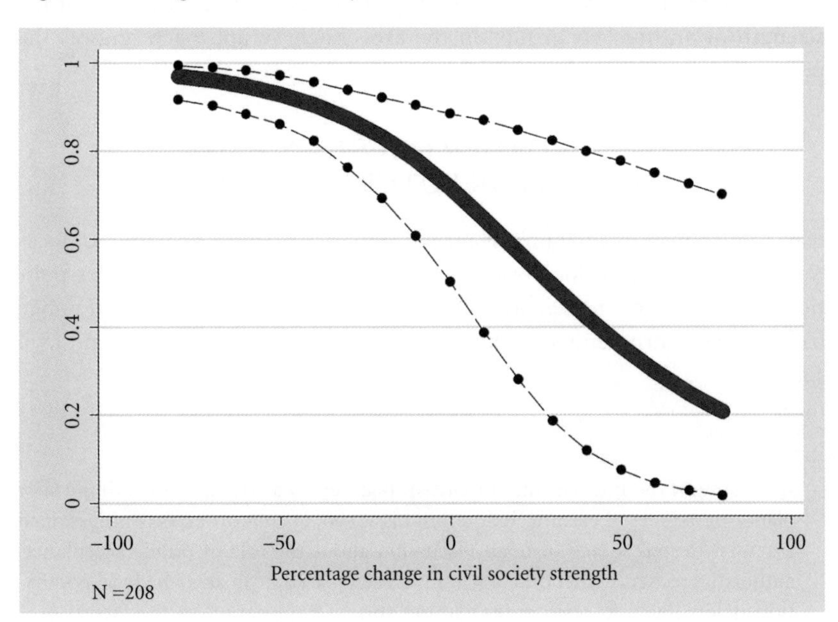

Note: Data from Aldrich, 2008a and Aldrich, 2008b

Planners in North America, Japan, France, and England who have taken civil society seriously have recognized an empirical fact: A strong local civil society can more easily sabotage unwanted projects than weak ones. Dense interpersonal connections facilitate the community's capacity to work together, mobilize, and defend themselves against projects they see as bringing primarily negatives in their backyards. Weaker communities, though, where individuals do not volunteer join groups, or work together easily are more likely to fragment under pressure and make better targets for developers.

Not all authorities seem to recognize the power of the people, so to speak, when it comes to potential siting conflict. The British Government in mid-2007, for example, sought to embark on a new nuclear program, but as experts have pointed out, "The availability of potential sites will therefore directly affect the government's view of the overall feasibility of a new nuclear build programme and the development of its energy policy" (Jackson & Jackson, 2006, p. 5). Many pages of a commissioned report on potential sites underscored the various technical criteria for siting plants: distance from existing electrical grids, need to be near supplies of water for cooling, and distance from highly populated areas to allow for evacuation. But nowhere did the report discuss local reactions. Despite various graphs and figures in the report that discussed the appropriateness of new and existing nuclear sites, none of the facts involved local opinion polls, measures of civil society, or the strength of antinuclear groups in the area. Such an approach ignores the power of civil society at its peril.[6]

BROADER QUESTIONS: AVENUES FOR FUTURE RESEARCH

Having demonstrated the importance of civil society in the heuristics of locational planning, I now look briefly to broader issues relevant to NIMBY politics. I envision two larger areas of theory growing out of work on local public bads: policy instruments and the benefits of conflict between state and civil society.

---

[6] Interestingly, the bottom line for British planners is a strong push to build new plants on the sites of existing, but decommissioned, or soon-to-be-decommissioned reactors. Despite a lack of open discussion about the role of public acceptance, authorities recognize that once a plant has been placed in an area, the local population is less likely to resist future sitings due to habituation to the facility (see Aldrich, 2008a).

First, a relatively uncharted field of study focuses on the policy instruments used both by the state and by social movements in conflict over these types of issues. When seeking to convince recalcitrant local communities in Japan of the benefits of hosting nuclear power plants, despite any accompanying risks, the government has created a number of new policy tools to assist its goals. Among these are "submerged" taxes on electricity use that funnel money to peripheral communities that agree to act as host communities. The *Dengen Sanpō*, or Three Power Source Development Laws, created in 1974 by the Japanese government, can provide often-impoverished, depopulated local rural communities in Japan with up to $20 million a year in subsidies, grants, and special programs. France and Japan both have experimented with similar jet-fuel and travellers' taxes that funnel money to communities hosting airports; the bulk of the funds are spent on noise mitigation strategies.

Beyond important questions of such submerged, or "iceberg," policy instruments that impact the attitudes of local residents without being affected by typical budgetary or legislative politics (Howard, 1997; Hacker, 2000), these tools are only some in a broader toolkit of potential instruments available to decision makers. In the United States, the 2005 *Kelo v. New London* decision has reinvigorated discussions of the takings clause and the scope of the government's power of eminent domain. The three-decade struggle over the Narita Airport was touched off by the decision to use land expropriation against local farmers (Apter & Sawa, 1984). Research on policy instruments is being stimulated by a new generation of scholarship. The January 2007 issue of *Governance: An International Journal of Policy, Administration, and Institutions* brought in the guest editors Pierre Lascoume and Patrick le Gales to reexamine the issue of policy instruments.[7] Our work on cases of state–civil society interaction over public bads can contribute a great deal to answering questions about policy instruments.

A second field of research to which our studies can contribute is that of the *benefits of conflict* between civil society, whether local or extralocal, and authorities, whether governmental or private. Some earlier researchers worried about "excessive" conflict between state and civil society that might indicate long-term instability or potential collapse of industrialized governments (Crozier, Huntington, & Watanuki, 1975). Despite such concerns, they seem to have been misplaced. Research has now focused on the ways in which conflict between even local residents and states can lead to better, more sustainable plans.

---

[7] See also Aldrich (2005a) for a discussion of country-specific tools when siting controversial facilities.

For example, McAvoy (1999) emphasized how contestation between communities in Minnesota and the state and federal governments over planned waste-disposal facilities forced the state to reconsider its options. Looking back on the options available to the state at the time, it is now clear that had the state gone ahead, it would have invested millions of dollars in poorly tested, less-useful technologies and been locked into a less efficient approach to waste. Instead, "Not in My BackYard" opposition from citizens pushed the state to create new, more efficient plans for handling waste.

Similarly, Aldrich (2008a) has illuminated how long-term contestation between Japanese citizens and the central government over the siting of nuclear power plants forced the development of a new set of "softer" policy instruments that moved away from expropriation, police force, and other standard, Weberian policy tools. While a number of government agencies in Japan handled conflict with local citizens over the Postwar period, only those bureaus encountering long-term, high-level contestation from well-organized civil society developed more sustainable tools for handling conflict now and in the future. Hence contentious interaction between states and civil society in the field of unwanted projects can create new policy instruments and innovative public policies, and we should use our growing body of knowledge to further such investigations.

TOWARDS THE FUTURE

How can we apply our growing understanding of the role of civil society in facility siting to larger, worldwide concerns? Given new recognition of the dangers of global warming and greenhouse gas emissions (thanks to new media such as Al Gore's *An Inconvenient Truth*), a number of technologies—both familiar and untested—are being raised as potential solutions to reducing emissions. The nuclear-power industry is gearing up for a "renaissance" of nuclear power, wind turbines are becoming popular among fans of green energy, and proponents of carbon sequestration hope to tackle the back end of the problem by placing carbon dioxide in buried reservoirs.

In this new environment, with planners concerned over the gasses emitted by typical thermal-type electricity generation, such as coal, nuclear power has been given a new life. The Bush Administration, through a series of generous tax breaks and subsidies, has sought to restart the North American commercial nuclear power program, and some investors have joined the bandwagon. Proponents of nuclear power have adopted a new "frame" when discussing it, emphasizing the green aspects of nuclear technology despite arguments beginning in the 1970s from antinuclear groups that such power plants are the cause of, not the solution to, environmental problems (Pralle, 2007).

Whatever the future holds for nuclear power and alternative energies, the fate of such facilities rests solely in the hands of local residents. Environmental groups, local fishing cooperatives, and residents already oppose trials of carbon sequestration and research suggests little difference between plans for sequestration and other projects promoting the "public good" (Schively, 2007b). Residents in upstate New York have complained about the activities of wind companies seeking to construct new power-generating turbines (Confessore, 2008). While policy makers may envision the smooth implementation of new technologies, they often forget to weigh the potential costs of these projects on local residents. Civil society has been a critical factor in determining the locations for and ways by which divisive facilities are sited in the past, and it will continue to play a role around the world as citizen expectations for their governments rise and transparency increases. Too often, planners and social scientists, alike, have overlooked the fact that the power—and hence the future success or failure of these projects and technologies—truly lies in the hands of the people.

REFERENCES

Abel, T. (2001). *Community involvement in environmental justice decision making*. Paper prepared for the Annual Meeting of the Midwest Political Science Association, Chicago.

Aldrich, D. P. (2005a). The limits of flexible and adaptive institutions: The Japanese government's role in nuclear power plant siting over the post war period. In S. H. Lesbirel & D. Shaw (Eds.), *Managing conflict in facility siting* (111–136). Surrey, UK: Edward Elgar Publishers.

Aldrich, D. P. (2005b). Controversial facility siting: Bureaucratic flexibility and adaptation. *The Journal of Comparative Politics*, *38*(1), 103–123.

Aldrich, D. P. (2008a). *Site fights: Divisive facilities and civil society in Japan and the West*. Ithaca, NY and London: Cornell University Press.

Aldrich, D. P. (2008b). Location, location, location: Selecting sites for controversial facilities. *Singapore Economic Review*, *53*(1), 145–172.

Aldrich, D. P., & Crook, K. (2008). Civil society as a double-edged sword: Siting trailers in post-Katrina New Orleans. *Political Research Quarterly*, *61*(3), 379–389.

Altshuler, A., & Luberoff, D. (2003). *Mega-projects: The changing politics of urban public investment*. Washington, DC: Brookings Institution Press.

Apter, D., & Sawa, N. (1984). *Against the state: Politics and social protest in Japan*. Cambridge, MA: Harvard University Press.

Becker, R. A. (2004). Pollution abatement expenditure by U.S. manufacturing plants: Do community characteristics matter? *Contributions to Economic Analysis & Policy*, *3*(2), 1–21.

Brion, D. (1991). *Essential industry and the NIMBY phenomenon*. New York: Quorum Books.

Broadbent, J. (1998). *Environmental Politics in Japan: Networks of Power and Protest*. Cambridge: Cambridge University Press.

Bullard, R. (1994). Overcoming racism in environmental decision making. *Environment*, *36*(4), 10–17.

Bullard, R. (2000). *Dumping in Dixie: Race, class, and environmental quality*. Boulder, CO: Westview Press.

Cho, W. K., & Gimpel, J. G. (2007, August 30–September 1). *The spatial and temporal distribution of capital and labor in an election campaign*. Paper presented at the Annual Meeting of the American Political Science Association, Chicago.

Clingermayer, J. (1994). Electoral representation, zoning politics, and the exclusion of group homes. *Political Research Quarterly*, *47*(4), 969–984.

Cole, L. W., & Foster, S. R. (2001). *From the ground up: Environmental racism and the rise of the environmental justice movement*. New York: New York University Press.

Confessore, N. (2008, August 17). In rural New York, windmills can bring whiff of corruption. *The New York Times*.

Crozier, M., Huntington, S., & Watanuki, J. (1975). *The crisis of democracy: Report on the governability of democracies to the trilateral commission*. New York: New York University Press.

Ehrman, R. (1990). *NIMBYism: The disease and the cure.* London: Centre for Policy Studies.

Ethington, P. J., & McDaniel, J. A. (2007). Political spaces and institutional places: The intersection of political science and political geography. *Annual Review of Political Science, 10,* 127–142.

Falk, J. (1982). *Global fission: The battle over nuclear power.* New York: Oxford University Press.

Feldman, E. (1985). *Concorde and dissent: Explaining high technology project failures in Britain and France.* Cambridge: Cambridge University Press.

Feldman, E., & Milch, J. (1982). *Technocracy versus democracy: The comparative politics of international airports.* Boston: Auburn House Publishing.

Frey, B., Oberholzer-Gee, F., & Eichenberger, R. (1996). The old lady visits your backyard: A tale of morals and markets. *Journal of Political Economy, 104*(6), 1297–1313.

Garcia-Gorena, V. (1999). *Mothers and the Mexican antinuclear power movement.* Tucson, AZ: University of Arizona Press.

Groth, D. (1987). Biting the bullet: The politics of grass-roots protest. Unpublished PhD dissertation. Stanford University, Stanford.

Hacker, J. (2000). Boundary wars: The political struggle over public and private social benefits in the United States. Unpublished dissertation. Yale University, New Haven.

Hagiwara, Y. (1996). *Yanba Dam no tatakai* [The struggle over Yanba Dam]. Tokyo: Iwanami Shoten.

Hamilton, J. (1993). Politics and social costs: Estimating the impact of collective action on hazardous waste facilities. *RAND Journal of Economics, 24*(1), 101–125.

Hasegawa, K. (2004). *Constructing civil society in Japan: Voices of environmental movements.* Melbourne, Australia: TransPacific Press.

Howard, C. (1997). *The hidden welfare state: Tax expenditures and social policy in the United States.* Princeton, NJ: Princeton University Press.

Hoyman, M. (2001). Prisons in North Carolina: Are they a viable strategy for rural communities? *International Journal of Economic Development*, S.P.A.E., special volume on Community Economic Development *4*(1), 1–35.

Hoyman, M., & Weinberg, M. (2006). The process of policy innovation: Prisons as rural economic development. *Policy Studies Journal, 34*(1), 95–112.

Hubbard, P. (2005). Accommodating otherness: Anti-asylum centre protest and the maintenance of white privilege. *Transactions of the Institute of British Geographers, 30*(1), 52–65.

Hurley, A. (1995). *Environmental inequalities: Class, race, and industrial pollution in Gary Indiana, 1945–1980.* Chapel Hill, NC: University of North Carolina Press.

Inhaber, H. (1998). *Slaying the NIMBY dragon.* New Brunswick, NJ: Transaction Publishers.

Inhaber, H. (2001). NIMBY and LULU. *Cato Review of Business and Government, 14*(4), 1–4.

Jackson, I., & Jackson, S. (2006). Siting new nuclear power stations: Availability and options for government. Discussion Paper for DTI Expert Group. UK: Jackson Consulting Limited.

Jenkins-Smith, H., & Bassett, G. (1994). Perceived risk and uncertainty of nuclear waste. *Risk Analysis, 14*(5), 851–856.

Kunreuther, H., & Kleindorfer, P. (May 1986). A sealed bid auction mechanism for siting noxious facilities. *The American Economic Review, 76*(2), 295–299.

Lee, M., & Bartkowski, J. (2004). Love thy neighbor? Moral communities, civic engagement, and juvenile homicide in rural areas. *Social Forces, 82*(3), 1001–1035.

Lesbirel, S. H. (1998). *NIMBY politics in Japan: Energy siting and the management of environmental conflict.* Ithaca, NY: Cornell University Press.

Levy, J. (1999). *Tocqueville's revenge: State, society, and economy in contemporary France.* Cambridge, MA: Harvard University Press.

McAvoy, G. (1999). *Controlling technocracy: Citizen rationality and the NIMBY syndrome.* Washington, DC: Georgetown University Press.

McKean, M. (1981). *Environmental protest and citizen politics in Japan.* Berkeley, CA: University of California Press.

Mitchell, R., & Carson, R. (1986). Property rights, protest, and the siting of hazardous waste facilities. *The American Economic Review, 76*(2), 285–290.

Mohai, P., & Bryant, B. (1992). Environmental racism: Reviewing the evidence. In B. Brant & P. Mohai (Eds.), *Race and the incidence of environmental hazards: A time for discourse* (163–176). Boulder, CO: Westview Press.

Munton, D. (1996). Siting hazardous waste facilities, Japanese style. In D. Munton (Ed.), *Hazardous waste siting and democratic choice* (181–229). Washington, DC: Georgetown University Press.

Nakagawa, Y., & Shaw, R. (2004). Social capital: A missing link to disaster recovery. *International Journal of Mass Emergencies and Disasters, 22*(1), 5–34.

Nelkin, D., & Pollak, M. (1981). *The atom besieged.* Cambridge, MA: MIT Press.

Olson, M. (1965). *The logic of collective action: Public goods and the theory of groups.* Cambridge, MA: Harvard University Press.

Pine, J., Marx, B., & Lakshmanan, A. (2002). An examination of accidental-release scenarios from chemical processing sites: The relation of race to distance. *Social Science Quarterly, 83*(1), 317–331.

Poley, L., & Stephenson, M. (2007). Community and the habits of democratic citizenship: An investigation into civic engagement, social capital and democratic capacity-building in U.S. cohousing neighborhoods. Paper prepared for the American Political Science Association annual meeting, Chicago.

Pralle, S. B. (2007). Framing trade-offs: The politics of nuclear power, dams, and wind energy in the age of global climate change. Paper prepared for the American Political Science Association annual meeting, Chicago.

Putnam, R. (1993). *Making democracy work. Civic traditions in modern Italy.* Princeton: Princeton University Press.

Putnam, R. (1995). Bowling alone: America's declining social capital. *Journal of Democracy, 6*(1), 65–78.

Putnam, R. (2000). *Bowling alone: The collapse and revival of American community.* New York: Simon & Schuster.

Putnam, R. (2007). E pluribus unum: Diversity and community in the twenty-first century; The 2006 Johan Skytte Prize Lecture. *Scandinavian Political Studies, 30*(2), 137–174.

Quah, E., & Tan, K. C. (2002). *Siting environmentally unwanted facilities: Risks, trade-offs, and choices.* Northampton, MA: E. Elgar.

Rabe, B. (1994). *Beyond NIMBY.* Washington, DC: Brookings Institution.

Ramseyer, J. M., & Rosenbluth, F. (1993). *Japan's political marketplace.* Cambridge, MA: Harvard University Press.

Rüdig, W. (1994). Maintaining a low profile: The anti-nuclear movement and the British State. In H. Flam (Ed.), *States and anti-nuclear movements* (70–100). Edinburgh: Edinburgh University Press.

Schively, C. (2007a). Understanding the NIMBY and LULU phenomena: Reassessing our knowledge base and informing future research. *Journal of Planning Literature, 21*(3), 255–266.

Schively, C. (2007b). Siting geologic sequestration: Problems and prospects. In E. Wilson & D. Gerard (Eds.), *Carbon capture and sequestration: Integrating technology, monitoring, and regulation.* Ames, IA: Blackwell Publishing.

Shemtov, R. (2003). Social networks and sustained activism in local NIMBY campaigns. *Sociological Forum, 18*(2), 215–244.

Sherman, D. (2006). Not here, not there, not anywhere: The federal, state, and local politics of low-level radioactive waste disposal in the United States. Paper prepared for the Northeastern Political Science Association annual meeting, Boston.

Smith, H., & Kunreuther, H. (2001). Mitigation and benefits measures as policy tools for siting potentially hazardous facilities: Determinants of effectiveness and appropriateness. *Risk analysis, 21*(2), 371–382.

Smith, S. (Ed.). (2000). *Local voices, national issues: The impact of local initiative in Japanese policy-making.* Center for Japanese Studies, University of Michigan Monographs in Japanese Studies. Ann Arbor, MI.

Takahashi, L. (1998). *Homelessness, AIDS, and stigmatization: The NIMBY syndrome in the United States at the end of the twentieth century.* Oxford: Clarendon Press.

Tanaka, K. (1972). *Building a new Japan: A plan for remodeling the Japanese archipelago.* Tokyo: Simul Press.

Touraine, A., Hegedus, Z., Dubet, F., & Wieviroka, M. (1983). In P. Fawcett (Trans.), *Anti-nuclear protest: the opposition to nuclear energy in France.* Cambridge: Cambridge University Press.

United Church of Christ. (1987). *Toxic wastes and race: A national report on the racial and socioeconomic characteristics of communities with hazardous waste sites.* New York: UCC Commission for Racial Justice.

van der Horst, D. (2007). NIMBY or not? Exploring the relevance of location and the politics of voiced opinions in renewable energy siting controversies. *Energy Policy, 35*(5), 2705–2714.

Weingart, J. (2001). Waste is a terrible thing to mind: Risk, radiation, and distrust of government. Princeton: Center for Analysis of Public Issues.

Wellock, T. (1978). *Critical masses: Opposition to nuclear power in California, 1958–1978*. Madison, WI: University of Wisconsin Press.

Williams, W., & Whitcomb, R. (2007). *Cape wind: Money, celebrity, class, politics, and the battle for our energy future on Nantucket Sound*. New York: PublicAffairs.

Wolverton, A. (2002, July). *Does race matter? An examination of a polluting plants location decision*. Draft Copy, National Center for Environmental Economics, U.S. Environmental Protection Agency, Washington, DC.

7

Site Selection of LULU Facilities: The Experience of Taiwan

Chang-tay Chiou

INTRODUCTION

LULU (Locally Unwanted Land Uses) syndrome has become one of society's controversial issues. No matter where it occurs—in a developed country or in an underdeveloped one—proposed construction of new facilities is often met with forceful public opposition. Related examples are landfills or solid waste incinerators, airports, prisons, low-income housing projects, electric power stations, transportation facilities, recreational facilities, water supply facilities, social service facilities, etc. The situating of LULU facilities is subject to criticism by community residents, concerned grassroots' groups, and local LULU politicians. Site selection protests have become frequent, and some have even turned violent. NIMBY (Not In My BackYard) syndrome has grown as project opponents have attempted to alter, delay, or stop public construction projects. The site planner invariably faces some form of paralysis in the site selection of public facilities. Thus, breaking through the predicament of LULU syndrome in the site selection process becomes one of the most exigent tasks for site practitioners and urban planners.

Indeed, site selection of a public infrastructure decision and the NIMBY syndrome situation arising is a common scenario of industrialized nations' worldwide (Lake, 1987; Popper, 1987). NIMBYism research has always been the focus of urban planners and site practitioners. How can a LULU, NIMBY, or NIMTOO (Not In My Term Of Office) site be built without setting off riots or endless litigation? Are there any possibilities to effectively resolve the tension caused by annoying NIMBY constituencies? There are several related studies, but the answers are inconclusive. The aim of this research is to examine NIMBYism in Taiwan and offer our experiences to place LULU facilities successfully.

This paper focuses on Taiwan as an example, specifically the location selection and execution of facility construction. This is because it is not only one of the most active members of the four Asia "dragons," but is also a miracle of economic development. As an island country with a high population density, the Taiwanese are not only well educated but are more likely to pay particular attention to improving their environment and living standards compared to other developing countries. Antipollution movements are common in this transitional society. Therefore, public protest and social movements derived from NIMBY syndrome in Taiwan are more passionate than other developing countries.

The completion of this study is based on a systematic survey of 24 LULU cases in Taiwan, 9 cases of electric power stations, and 15 cases of solid waste incinerators, which were established over the last decade. Unlike past literature concerned with the effect of "economic" factors such as compensation or auctions in situating the LULU site, the author urges local government decision makers to be more sensitive to the influence of "noneconomic" factors in making the LULU decision, including public participation, social trust, and local politics in the site selection process. It is my hope that the Taiwan experience will give readers some useful lessons on how to promote a successful site location and plan for its smooth operation. Most importantly, I recommend some ways of resolving NIMBY confrontations, so policy planners can carry out their public infrastructure projects successfully and thereby reduce the cost burden on society.

REDEFINING "LULU SYNDROME"

There are many terms regarding LULU; for instance, Build Absolutely Nothing Anywhere Near Anything (BANANA), Not In My Bottom Line (NIMBL), Not On Our Street (NOOS), etc. According to NIMBY researchers, LULU syndrome began 30 years ago in the United States. Then, the focus of NIMBYism was on environmental pollution facilities, such as a waste processing plant (Halstead et al., 1993), toxic-treatment facilities (Bryant & Mohai, 1992), airport (Hall, 1980), etc. But recently many cases have occurred having nothing to do with pollution, such as human service facilities (Takahashi & Dear, 1997), and site selection of prisons in rural North Carolina (Hoyman & Weinberg, 2006; Sechrest, 1992).

What is NIMBY? The definition of NIMBY is confusing. Inhaber (1998) pointed out NIMBY syndrome is actually a kind of hatred or dislike of the public infrastructure's repeal consciousness. He describes such a phenomenon as a "dragon" and the process of resolution as "slaying the NIMBY dragon." As Richman and Boerner (2006, p. 37) indicated LULU is defined as socially desirable land use that broadly distributes benefits, yet is difficult or impossible to implement because of local opposition. Two important characteristics define LULU facilities. The first is the project will generate an overall increase in social surplus. Second, the nature of the costs and benefits associated with these facilities virtually assures local opposition. Lake (1993) observed LULU occurs because of two factors: (1) The public facility itself is required in some manner and provides important social benefits and (2) local parochialism is an obstacle to a practical and social good. Therefore, LULU syndrome is a tough bottleneck for environmental planners to break through in planning and executing public facilities. Such gridlock cannot be settled by any rational or technical means.

Dear defines NIMBY as "the motivation of residents who want to protect their turf. More formally, NIMBY refers to the protectionist attitudes of land oppositional tactics adopted by the community groups facing an unwelcome development in their neighbourhoods. Such controversial developments encompass a wide range of land use proposals" (1992, p. 288). Vittes et al. (1993) recognized NIMBY as (1) a passive attitude towards and/or denial of the public infrastructure considered harmful to the community's right to living and environmental well-being; (2) the syndrome is basically environmentalism in that it stresses the value of the environmental effects of the standard of living of those living near the public infrastructure; and (3) NIMBY syndrome is actually sometimes not based on technical, economical or administrative knowledge or discussion; rather, it is a rejection phenomenon.

Although LULU attitudes are sometimes regarded as "rights" of the individuals, many studies have shown that LULU is a selfish, egocentric, and political trait (Hunter & Leyden, 1995). Common elements of LULU protesters, who subverted a well-conceived and essential disposal facility, include fear, meetings, emotions, and politics. It is always hard to dissuade them through rational discussion. Many developed countries look at it as an obstacle to public infrastructure construction. In the United States, for example, from 1980 to 1987, 81 toxic waste treatment plants were planned for construction . Only eight facilities were completed. The main obstacle was the LULU syndrome. Irrational arguments were made to obstruct public construction (Lake, 1987; O'Hare & Sanderson, 1988).[1]

[1] Through their research, O'Hare & Sanderson (1988) found in this instance of NIMBY only 2.4% really cared about the environmental effects and property value, 45.7% were apathetic or did not care, and 25.5% were concerned with their health and safety. Therefore, the content leading up to the NIMBY syndrome consisted of many irrational factors.

Based on the above description, the author found LULU syndrome has the following four distinct characteristics: (1) the benefits of the public infrastructure are shared by the whole society at the expense of those local residents affected, (2) it was determined the degree of acceptance by local residents varied with the distance between the public facility and their homes, (3) the LULU syndrome is an irrational response; most carry it out through passive but wilful resistance—or simply put, for the sake of opposition alone, and (4) initially, LULUs sometimes arose between the technical experts and government officials and the community-at-large over the value and purpose of the construction.

ECONOMIC VS. NONECONOMIC FACTORS CONTRIBUTING TO SITING LULU FACILITIES

How can LULU facilities be sited without causing neighbourhood opposition or resident protest? What factors contribute to a successful decision for a LULU site? Past literature has paid much attention to the importance of site location and the effect of economic incentives. Economic instruments, such as auctions or compensation strategies, must be implemented for the victims adjacent to the selected LULU site. Site proximity should cause irrational neighbourhood opposition. Compensation or public auction is the only effective way to solve the gridlock of the LULU syndrome. Accordingly, the study of LULU phenomena has overwhelmingly focused on the economic-incentive instruments and neighbourhood-demographic characteristics as predictors of hostility, with housing in low-income, transient, heterogeneous neighbourhoods expected to generate the least resistance (Zippay, 2007).

Lehr and Inhaber (2003, p. 402) suggest a creative "reverse Dutch auction" approach to effectively end the stalemate caused by LULU syndrome that so greatly retards progress of public construction facilities. The guidelines behind using the reverse Dutch auction to site LULU, such as waste-disposal sites or electric power stations, are (1) the site has to be volunteered and (2) environmental standards should not be reduced. Basically, there is no "best site." Although the reverse Dutch auction strategy has many advantages, Neuzil (2003) argues a LULU will have different effects on different members of the population. Landowners adjacent to a proffered site without any benefits from the auction must oppose this proposal. The underlying assumption of the reverse Dutch auction will fail because it is difficult to have a specific area that will be universally volunteered. How do we expect every citizen in the LULU area to accept the auction proposal without opposition? How do we reach the consensus of residents to make the decision of the LULU site? If through referendum, "the tyranny of majority" is another problem—the public interest of the minority might be sacrificed under the result of the referendum. Thus, it would not avoid litigation, political battles, and facility delays.

The market approach by Inhaber (1992) also illustrates those who cry "NIMBY" are telling us the cost, real or perceived, of a LULU in their vicinity is high to them. Thus, a public auction is suggested to deal with LULUs, money has to be paid to the community making a bid. Richman and Boerner (2006) develop a transaction cost economic model for regulation and apply this model to environmental site selection regulations designed to overcome NIMBY political opposition. However, how fair is fair enough? Many LULU cases have clearly confirmed the available money is not always enough to be equally distributed to every resident in the LULU area. In any society, equal distribution is hardly achieved by resource distributors. According to Dorshimer (1996), economic developers and others have had great difficulty in the past situating high-impact projects because they (1) failed to recognize and avoid the technical rational vs. culture rational root cause of the NIMBY phenomenon, (2) have not successfully presented and defined the benefits and costs to all affected parties, and lastly, (3) have failed to reach equitable and fair agreements on the redistribution of these benefits and costs.

In *Controlling Technocracy*, McAvoy (1999) applies the concept of NIMBY to hazardous waste-facility site selection in Minnesota. He found public participation, which is a noneconomic factor, has a positive impact on solving the predicament of the LULU facility. Residents are not always irrational, self-interested, and ill informed. After participating in public hearings and intensive communication, local people are highly knowledgeable about the LULU site facilities. McAvoy (1999) concluded democratic decision making may be cumbersome and slow, but can change the NIMBY syndrome into YIMBY (Yes In My BackYard). Thus, Waugh (2002) values public participation in site selection policy making. In fact, more and more site practitioners and urban planners are adopting the "community governance approach" instead of the "market approach." In this model the goal is to reach collaborative, consensus-based decisions. Government, business, community groups, and citizens work together; and leaders share power, working to enable others to decide issues. In short, the new model would redefine politics away from the capital "P" politics of elections and campaigns and voting and contributions to the small "p" politics of self-government. The challenge of the coming decade is for America's leaders and citizens to adapt themselves and their communities to this new world (Gates, 1999). For example, the solution envisioned by most site-selection agencies now is education. To carry out public education, public meetings are held in which the potential host communities are lectured by scientists and administrators. The experts usually explain the risks are very small and the chances of environmental damage from a well-engineered facility are almost negligible. Then, they sit back and wait for nods of agreement from the locals.

The community-governance mode implies the importance of democratization of environmental policy. Residents not only have the right to know how the LULU site selection facilities have been established around the neighbourhood, but also have the right to access information regarding siting decision making. As Gould (2002) and Korten (1995) pointed out, only direct democracy in LULU site selection facilities can prevent developmental and economic interests from subverting environmental protection efforts.

Sénécal and Reyburn (2006), using Montreal as an example, have constructed an integrated conceptual framework that includes encouraging citizens' participation, negotiating solutions, and considering planning practices, to deal with the dilemma of NIMBY. The central concern of the conceptual framework is clearly the noneconomic dimension of LULUs. Curic and Bunting urge site planners to achieve balance between public participation in the planning process and the renewed debate about what constitutes "good" urban form, and planning policies (2006, p. 219). Kurland (1992) believes risk communication is a crucial component of any successful venture involving public perceptions of high risk.

McAvoy (1999) challenged the stereotype of NIMBY forces as overly excited, parochial, selfish, and ill-informed. He found local residents may well have considerable knowledge of technical issues. For that reason, Minnesota officials expected local interests would oppose the experts' preferences and encouraged efforts to limit public input in the site-selection process.

The effect of the location/economic approach to find sites for LULUs is clearly limited. Everyone is in favour of finding a place for a LULU, as long as it is at least a hundred miles away from them. In a densely populated country such as Taiwan, official planners have limited choice for site selection. Therefore, they have to be more concerned with the significance of "noneconomic" instead of "economic" factors. Although money can reconcile the tension between the vendors and residents, it cannot get to the bottom of the LULU problem. By focusing on a biomedical incinerator in a small city, Dunn, North Carolina, Sellers argues that "for waste incinerators, fear of risk to health and safety is probably the most significant one." Others concerns include, for instance, "the concern over possible mismanagement of such facilities, falling property values, and the unfair burdening of one community with waste created statewide or by neighboring states" (1993, pp. 460–461).

As McAvoy (1999) mentioned, lack of participation in the local community in the facility site selection process is a key factor of NIMBY syndrome.[2] In fact, if we could involve and educate the local community to understand the technical aspects and the societal value, this would help bring all viewpoints to a compromise, allowing the construction of the public facility to proceed more smoothly.

As pointed out earlier, LULU syndrome arises from economic factors but is much more complex than a conglomeration of economic, social, psychological, or political responses by residents to local controversies (Takahashi, 1997). In a systematic study of the NIMBY phenomenon in Taiwan, this research summarized the NIMBY syndrome as exhibiting the following factors:

(A) Economic factors, including (1) health and property safety concerns. To the community residents, the most controversial cases of LULU syndrome occur when there is a profound effect on the emotional and psychological level and when there is damage to personal property. In the site selection of power plants and waste incinerators, this article focuses on how the following four specific risks affected the local community: health, safety, property value, and farmers' and fishermen's yield. (2) Compensation for environmental effects. No matter what rationale is given, there is no avoiding a confrontation between the facility owners and the local community residents, which creates a LULU syndrome. Therefore, a monetary settlement of some kind is unavoidable. The community will usually request compensation either for the local construction itself or a monetary or nonmonetary compensation for the health, property, and psychological losses of community residents. In Taiwan, environmental compensation has been recognized as a key means of settling the NIMBY syndrome. This article examines the following three items for analysis: Did the power plants or waste incinerators site selections improve the economic development? Did the power plants or waste incinerators site selections create jobs? What is the most compensation that would be asked of the

---

[2] McAvoy (1999) indicated NIMBY syndrome originally began in the United States. The increasing democratic involvement of local residents clashed with the advancement of technology in the public facility construction policy arena. Government policy relied heavily on the scientific and technical expertise of planners, while the local community was viewed as too emotional or irrational to be involved in policy decisions. This led to the NIMBY syndrome, a movement that has ceaselessly developed and grown as a major obstacle to public facility construction. Therefore, lack of participation in the facility site selection process becomes a key factor of NIMBY syndrome.

power plants or waste incinerators? (3) Living standard risk: The public facility construction, of course, affects the local residents' quality of life. Lee (1997) indicated the living environment includes the natural and the man-made environment. The latter includes the real and unreal environment. Based on his definition, the environmental living standard includes land uses, environmental hygiene, traffic impact, landscape, air pollution, water contamination, noise disturbances, etc. This article will also discuss the following questions. How will this site selection affect total living quality? What did residents consider to be the most unacceptable impact?

(B) Noneconomic factors, including (1) lack of public participation: LULU syndrome appears when expert opinion is relied on solely for government public policy, neglecting public opinion and civic engagement altogether. Therefore, to smoothly settle LULU syndrome, it is important for local residents to participate in the public facility site selection and planning process. This article discusses the following three questions: Did this facility site selection ever have a public hearing? Did the facility-site owner ever provide detailed information to the local community? Was this case suitable to provide a local referendum? (2) Credibility deficiency: Most public agencies and private firms responsible for waste management and facility-site selection in Canada and the United States suffer from low credibility. Their traditional approach to facility site selection has only exacerbated earlier distrust (Rabe, 1994, p. 163). The LULU syndrome is a transparent example of a serious credibility crisis between local groups and civic officials; specifically citing a general distrust of the government's fairness and the entrepreneur's willingness to deal with and collect the environmental pollution arising from the situation. This creates distrust among the groups. Chiou (2002) points out the credibility gap is one of the key issues to study. Therefore in this article, I tried to address the following two items: Did local residents believe the power plants or waste incinerators owners were sincere in seeking to benefit the community? Did the local residents believe the owner intends to resolve any environmental pollution created and settle the safety issues? (3) Local politics: to the local Taiwanese politician, the logical resolution to LULU is to seek the vote of support from the local community concerned with the facility construction. Since the motivation to get involved is driven by a desire to win support for an incumbent election, the local politician tends to support the residents no matter how irrational the case. In the ballot consideration, the local representatives or civic leaders must approach the public for its opinion, no matter how obscure the situation is. Therefore, voting is the political process's main axis in Taiwan's present political situation. In the political arena, this article mainly focuses on whether any underground or local factions were involved.

RESEARCH METHODS

The LULU facilities in this study are solid waste incinerators and electric power stations island-wide (Table 1). The research methods depended on two parts. (1) The in-depth interview: The author had to interview the representatives of the solid-waste incinerators and electric power stations and the neighbourhood of both sites. Fifteen solid-waste incinerators and nine power plants, covering the entire island from south to north were surveyed: Chang-Sen power station and Tsin-Tow station in northern Taiwan, Hai-Du station in central Taiwan, Mai-Lou station and Chia-Fai station in southern Taiwan, and Hoc-Ping station in eastern Taiwan. Solid-waste incinerators in Taipei, Taoyuan, Hsinchu, Taichung, and Kaioshung were intensively examined. The author also interviewed the station manager or the public affairs representative of the solid-waste incinerators and electric power stations. For the local people, the community leader or the self-appointed group representative of the community was interviewed. In total, 30 people in power plants and 60 people in waste incinerators were interviewed. All the interviews were recorded. Interview questions were chiefly based on the two types of influencing factors mentioned above. Each in-depth interview took about 30 minutes to an hour.

Structured telephone interviews were also conducted with 260 residents for electric power stations and 765 for solid-waste incinerators. The former sample is fairly representative at the confidence level of 95 percent, having a sampling error of ±6.07 percent. The latter sample is satisfactorily representative at the confidence level of 95 percent, having a sampling error of ±3.54 percent.

The research protocols were conducted by the National Taipei University's Research Center for Public Opinion and Election Studies. The Center employed the CATI (Computer-Assisted Telephone Interview) system to

Table 1. Type of LULU Facilities and Samples of Interviews

Types of LULU	Cases	Sample of telephone interview	Sample of in-depth interview
Solid-waste incinerators	15	765	60
Electric power stations	9	260	30

survey the residents nearby the power plant sites and solid waste incinerators. We used the phone directories of each local community to conduct phone interviews. The demographics of the interviewed sample in terms of sex, age, education background, and occupation are indicated in Table 2. This method of research was the first of its kind in LULU research in Taiwan, and it is believed the strength of this sampling methodology lies in its random selection process. The results were not only representative of the community but were also very meaningful to this study.

Table 2. Demographics of the Interviewed Sample

Variable	Classification	Electric power stations % (n)	Solid-waste incinerators % (n)
Sex	Male	65.0 (169)	49.2 (376)
	Female	35.0 (91)	50.8 (389)
Age	20–29	12.7 (33)	14.6 (112)
	30–39	18.1 (47)	23.4 (179)
	40–49	35.8 (93)	26.8 (205)
	50–59	20.0 (52)	21.8 (167)
	60 or over	13.1 (34)	12.8 (98)
Education	Illiterate	3.8 (10)	4.4 (34)
	Elementary	12.3 (32)	13.5 (103)
	Junior high	18.1 (47)	10.8 (83)
	High school	36.9 (96)	35.6 (272)
	College/university	26.5 (69)	31.8 (243)
	Master or doctor	1.2 (3)	2.5 (19)
Occupation	Farmer/fisherman	9.2 (24)	3.8 (29)
	Blue-collar worker	25.0 (65)	16.6 (127)
	Retail/vendor	12.3 (32)	13.7 (105)
	Service	11.5 (30)	11.9 (91)
	Civil/military	9.6 (25)	9.0 (69)
	Housewife	14.6 (38)	5.1 (39)
	Retiree	5.4 (14)	19.6 (150)
	Unemployed	4.6 (12)	7.7 (59)
	Others	7.7 (20)	8.0 (61)

EMPIRICAL RESULTS

The following are the results and analysis of our research:

The Existence of NIMBY Syndrome

First, this paper discusses how we determine if NIMBY syndrome actually exists. As indicated above, this research analyzed the existence and seriousness of NIMBY syndrome in constructing Taiwan's electric power stations and solid waste incinerators in the north, centre, and south of the island. We examined three indicators: (1) The "welcome" indicator: this describes the local reaction of the LULU facilities being sited in the community. (2) The "favour" indicator: this describes whether the local community favoured or disfavoured the site decision. (3) The "fairness" indicator: this describes whether the decision to site the LULU facilities in the local community was made fairly. NIMBY syndrome is likely to exist if the three questions above are answered in the negative. If all the above answers are positive, then NIMBY syndrome is unlikely to exist.

Based on the results of the interviews with factory representatives, we concluded NIMBY syndrome did exist in all cases. This was determined by two factors: (1) A total of 24 vendors applied to construct power plants and solid-waste incinerators, but only 10 were completed. The causes of abandonment certainly cannot be blamed on the confrontation from the local community; however, the NIMBY syndrome was the main reason. (2) For all the power plants and solid-waste incinerators, each located in different areas, the local community formed self-relief organizations to oppose the power plant and converted the pervasive NIMBY sentiment into active confrontation.

According to the survey results, in the "welcome" indicator category, 51.9 percent and 48.1 percent of the participants had an either "unwelcome" or a "very unwelcome" attitude, respectively, to the power stations and waste incinerators being set in their neighbourhood. In the "favour" indicator category, 58.1 percent and 57 percent of the residents either "did not favour" or "strongly opposed," respectively, the site selection of the power stations and waste incinerators in their neighbourhood. In the "fairness" indicator category, negative attitudes were lower than the above two indicators but still 38.8 percent and 40.6 percent of the residents described the action as "very unfair" or "unfair," respectively. This figure is still higher than the "fair" or "very fair" categories, which were 29.6 percent and 28.4 percent, respectively (Table 3). Therefore, all three indicators—"welcome," "favour," and "fairness"—showed negative results, thereby confirming LULU syndrome existed.

Table 3. Survey Results on the Existence of LULU Syndrome

1. The "welcome" indicator: Did you welcome the site or not?					
	Very unwelcome	Unwelcome	Welcome	Very welcome	Unknown
Power stations	21.9% (57)	30.0% (78)	28.1% (73)	2.7% (7)	17.3% (45)
Incinerators	17.9% (137)	30.2% (231)	30.6% (234)	2.2% (17)	19.1% (146)
2. The "favour" indicator: Did you favour the site or not?					
	Very unwelcome	Unwelcome	Welcome	Very welcome	Unknown
Power stations	25.4% (66)	32.7% (85)	19.6% (51)	3.1% (8)	19.2% (50)
Incinerators	22.2% (170)	34.8% (266)	24.3% (186)	1.3% (10)	17.4% (133)
3. The "fairness" indicator: Did you feel that it is fair or unfair for the site to be here?					
	Very unfair	Unfair	Fair	Very fair	Unknown
Power stations	13.8% (36)	25.0% (65)	26.5% (69)	3.1% (8)	31.5% (82)
Incinerators	16.2% (124)	24.4% (187)	26.8% (205)	1.6% (12)	31% (237)

Economic Factors Influencing LULUs

• *Health and Property Concerns*

According to in-depth interviews with the residents concerned, it was found the power plant's site selection affected the community by the health and property concerns as follows: (1) people worried that the electromagnetic field would affect their health, especially in increasing the risk of cancer; (2) people were anxious they would die from electric shock; they were especially concerned for their children's safety; (3) they believed the presence of a high-voltage electric tower in the surrounding area would result in property value depreciation; (4) they were concerned the magnetic field would damage their agricultural produce. With regard to the solid-waste incinerators, the residents of the community are mainly afraid of air pollution (especially dioxin), water, and solid water pollution.

According to the survey results, 41.9 percent and 44.7 percent responded the power plants and solid-waste incinerators would have a negative effect on their health; 46.9 percent and 47.3 percent of the residents in power stations and solid waste incinerators said it would have no effect; 51.9 percent and 47.2 percent of the participants in the sites of power plants and solid-waste incinerators said it would have a negative effect on the property, and 38.8 percent and 37.4 percent thought there would be no effect, respectively. Of the power station residents, 47.3 percent and 33 percent of residents of the site of solid waste incinerators said it would impact agricultural produce and production; 31.2 percent and 26.9 percent thought there would be no effect (Table 4). From these results, we found the interviewed opinions were quite varied. Regarding the health and safety risks, the residents acknowledged the establishment of an electric power plant would carry low risk. However, they felt there was a great risk to real-estate values and agricultural production. Therefore, we could conclude the residents' antagonistic attitude towards the power plant was chiefly due to their property values rather than their health.

Table 4. Survey Results on Health and Property Concerns

	Many ill effects	Some ill effects	Slight ill effects	No ill effects at all	Unknown
1. The site selection: Would it affect your health or not?					
Power stations	14.2% (37)	27.7% (72)	36.5% (95)	10.4% (27)	11.2% (29)
Incinerators	19.3% (148)	25.4% (194)	36.7% (281)	10.6% (81)	8.0% (61)
2. The site selection: Does it affect your property value?					
Power stations	21.5% (56)	30.4% (79)	27.3% (71)	11.5% (30)	9.2% (24)
Incinerators	23.4% (179)	23.8% (182)	25.6% (196)	11.8% (90)	15.4% (118)
3. The site selection: Does it affect your agricultural production?					
Power stations	15.4% (40)	31.9% (83)	23.1% (60)	8.1% (21)	21.5% (56)
Incinerators	14.6% (112)	18.4% (141)	18.0% (138)	8.9% (68)	40.0% (306)

COMPENSATION OF ENVIRONMENTAL EFFECTS

According to Taiwan's experience, managing environmental effects on residents' health and property is termed "compensation." Two types of compensation are identified in Taiwan, that is, monetary or nonmonetary compensation. Where does the funding source for the compensation come from? Most cases are mandated by compensation law under the jurisdiction of the environmental agency; however, in some cases, it is a vendor's voluntary action that is unrelated to public authority.

According to the survey, there are currently two types of compensation to resolve NIMBY syndrome:

(1) Monetary compensation: The amount of mandatory compensation to landowners in the area surrounding the high voltage power tower would be

set by the government decree, plus 10 percent. There are two types of voluntary compensation. One is the compensation set aside from the construction company at the outset of the construction. The other is derived from the power-plant management for ongoing damage. In the former, the compensation covers ill effects incurred during the site selection to the construction process. The minimum amount is US$100 per person, and the maximum is US$600 per person. Voluntary compensation from managing the power plants was usually negotiated between the community and the vendor. For example, in one case, the power plant's compensation was taxing about 0.5 percent of US$1 million revenue. There is also a compensation for low-income families, the elderly, and the handicapped in the community. According to Chinese custom, in the mid-Autumn Moon Festival and Dragon Festival and the Chinese New Year, a small monetary compensation (or welfare tip) would be distributed to these individuals. Regarding compensation for solid-waste incinerators, an incinerator in central Taiwan was granted US$5 million for the environmental pollution and the loss of property. In addition, US$10 per ton for sold waste and garbage managed by the incinerators has to be paid to the community around the LULU sites. In total, the community earned US$2.8 million per year.

(2) Nonmonetary compensation: An example would be the funding of scholarships; job opportunities for the community, especially for blue collar workers with no special skills and temporary workers; donations to the local houses of worship or community activities; participation in recreational activities; or improvement projects to beautify and enhance the local environment. This nonmonetary compensation was given voluntarily (not mandated by law). These favours depended upon the goodwill between the vendor and the community.

How did the community respond to the vendor providing so many kinds of compensation? According to the report, 54.6 percent of respondents asked about the power stations and 73.2 percent of respondents asked about the incinerators—even with the entire donations—still believed the plant itself had not improved the economic development of the community. However, 60.8 percent and 73.6 percent of the participants thought the power plant and incinerators did not increase job opportunities, respectively (Table 5). The most common form of compensation local residents require is a reduction in the electricity bill (31.2 percent for power stations and 34.6 percent for incinerators); the second-most common was the construction of public-welfare facilities, such as parks and other recreational facilities (25 percent and 17.9 percent) for electric power stations and solid-waste incinerators, respectively; and others asked for direct monetary compensation (13.8 percent for power stations and 11.3 percent for incinerators). Therefore, in this case study, reducing residents' electric bills was the major consideration for mitigating confrontation (Table 6). Following siting of the LULU facilities, only 15 percent of the participants in the power station sites and 18.8 percent of the participants in the site of solid waste incinerators were satisfied with the compensation. A high proportion, 65.4 percent and 57.5 percent, respectively, of the residents did not have any comment on the satisfaction of compensation. It seems the residents around the LULUs do not have a positive opinion of the compensation by the vendors (Table 5).

Table 5. Survey Results on Environmental Compensation

1. The site selection: Does it help local economic development?					
	Not at all	Does not help	Helpful	Very helpful	Unknown
Power stations	25.4% (66)	29.2% (76)	36.5% (95)	3.8% (10)	5.0% (13)
Incinerators	37.8% (289)	35.4% (271)	16.7% (128)	1.4% (11)	8.6% (66)
2. The site selection: Does it help create jobs?					
Power stations	25.4% (66)	35.4% (92)	28.5% (74)	2.3% (6)	8.5% (22)
Incinerators	41.8% (320)	31.8% (243)	14.1% (108)	1.8% (14)	10.5% (80)
3. The site selection: Are you satisfied with your compensation?					
	Highly dissatisfied	Dissatisfied	Satisfied	Highly satisfied	Unknown
Power stations	7.7% (20)	11.9% (31)	13.8% (36)	1.2% (3)	65.4% (170)
Incinerators	9.2% (70)	14.5% (111)	17.6% (135)	1.2% (9)	57.5% (440)

Table 6. The Most Favourable Compensation Methods

Types of compensation	Power stations	Incinerators
Cash	13.8	11.3
Offer job opportunities	7.3	10.4
Support public facilities	25.0	17.9
Omit electric bill	31.2	34.6
Give scholarships	1.9	7.9
Community activities	1.5	6.8
Recipients for public assistance	7.3	11.0

Quality of Life Concerns

The following five risks were cited as the most important living standard concerns of the residents. (1) Too many high-voltage power towers. Natural gas–generated power plants used too many power lines, which affect the scenic environment. (2) Within the construction period, increased traffic congestion created more exhaust fumes and dust, resulting in mud when it rained. (3) Coal-derived power plants and solid-waste incinerators created air pollution. (4) Some of the equipment the power plants and solid-waste incinerators used was loud, which affected the peace and quiet in the community. (5) Waste water from the power plant and solid-waste incinerators would pollute the local rivers or waterways. According to the survey results, 58.9 percent of the participants recognized these ill effects would have an unacceptable effect on the quality of life. Within this figure, 54.8 percent considered air pollution to be the most intolerable, followed by water pollution (16.5 percent) and electromagnetic-field pollution (12.9 percent) (Table 7).

Table 7. The Most Unacceptable Types of Pollution

Types of pollution	Percent
Electromagnetic-field pollution	12.9
Noise and vibration pollution	6.6
Scenic pollution	2.8
Water pollution	16.5
Air pollution	54.8
Others	6.3

Noneconomic Factors Influencing LULUs

- *Lack of Public Participation*

The Taiwan government enacted the Environmental Impact Assessment Act in 1994 (EIA of 1994). The EIA of 1994 is adapted from the National Environmental Policy Act of 1970 of the United States. The EIA requires two steps in evaluating the environmental effects in building power plants. It requires the contractor to sponsor a meeting of all concerned parties to participate in a hearing and to disclose detailed information on the plant construction for the residents' feedback. If the community has any strong opinions, the vendor has to explicate his process and suggest improvements. To make the hearing pass smoothly, vendors had to keep things ambiguous, obfuscating both the language and the process. For example, to avoid confrontation on some limited factors, vendors would intentionally neglect to publish the meeting time/date and location, selectively send out invitations, or cause the hearing to become a propaganda meeting focused on convincing the local community of their objectives. Clearly, this was not the compromise sought between the vendor and the local community because it limited the opportunity for residents to oppose perceived the ill effects. Knowing this, people refused to participate in such meetings. Instead, they chose a direct form of confrontation, such as a demonstration to block plant construction.

We provided a questionnaire to survey the opinions of the local community. The results were: (a) 59.2 percent and 79 percent in the sites of power station and incinerators, respectively, were never notified of the hearing or participated in the hearing; (b) 95.4 percent and 97.3 percent of local people indicated the vendors did not have detailed information or only a little bit of information was provided, 4.6 percent and 2.7 percent, respectively, for power stations and incinerators. From this survey, we understand the vendor-sponsored hearing did not encourage community participation and the vendor did not provide comprehensive details on the plant plan. In taking such authoritarian actions, the question is: Did the vendor mean to carry out the referendum? Of the participants, 46.6 percent and 55 percent accepted the policy for power stations and incinerators, and 38.5 percent and 33.9 percent were opposed to the policy, respectively. When we further asked those opposed why they rejected the policy, their answers were they thought the power plant and incinerators construction required specialized knowledge and therefore indicated they did not have the ability to decide (Table 8).

- *The Credibility Gap Problem*

Trust is the foundation of social capital. Social capital is an invisible liability. Society's mutual activity forms this invisible liability. There are three dimensions to this: social network, social trust, and social boundaries. A society high in social capital is marked by considerable trust among its people. The participants of this society mutually understand and share the concept of boundaries, and create a social network of relationships around these principles.

Putnam (1993, 2000) acknowledges social capital is a unique characteristic of social organization. Trust, boundaries, and network—these can be promoted and compromised to elevate the efficiency of a society. Therefore, the higher the peoples' and vendor's mutual trust, the higher the social capital. With social capital, the vendor can smoothly execute the site selection and construction. According to our survey for the respondents' trust, (a) 43.8

Table 8. Survey Results on Lack of Public Participation in Site Selection

1. The site selection: Did they have a public hearing?					
	Many times	A few times	Only one time	Never knew about it	Unknown
Power stations	6.5% (17)	25.4% (66)	8.8% (23)	59.2% (154)	0
Incinerators	5.1% (39)	12.8% (98)	3.1% (24)	79% (604)	0
2. The site selection: Did they provide detailed information?					
	None	Only a little information	A moderate amount of information	Plenty of information	Unknown
Power stations	30.8% (80)	26.9% (70)	26.5% (69)	4.6% (12)	11.2% (29)
Incinerators	37.4% (286)	18.7% (143)	19.3% (148)	2.7% (21)	21.8% (167)
3. The site selection: Did you agree to have a referendum to decide the site selection?					
	Strongly disagree	Disagree	Agree	Strongly agree	Unknown
Power stations	11.2% (29)	27.3% (71)	31.2% (81)	15.4% (40)	15% (39)
Incinerators	13.1% (100)	20.8% (159)	29% (222)	26% (199)	11.1% (85)

percent and 55.9 percent of the participants in the sites of LULUs did not trust that the power plant and incinerators proponents had a sincere attitude towards improving the welfare or prosperity of the community. This figure was equal to the number of participants trusting the vendor; (b) 37.3 percent and 34.1 percent of the participants in the LULU sites did not believe the vendor had a sincere attitude towards improving the pollution and safety issues; 45.4 percent and 47.5 percent thought the result was mixed. This confused constituents. However, on the improvement of environmental pollution and safety threats, more respondents trusted the power plant. When we further interviewed the people's opinion on other long-term effects, we found the pollution created by the power plant was less serious than the pollution created by a polluted factory. Therefore, the people were more inclined to believe the power plant and incinerators would be more likely to improve the pollution effects and create a safe environment (Table 8). When we asked the vendors whether they thought the people's confrontation was rational, we found the vendor trusted the local community less on this issue. Invariably, they thought the local community was selfish and egocentric and did not believe the residents could take a rational attitude toward settling confrontations concerning the power plant site selection. From the above-mentioned, the local community and vendor had a serious "credibility gap." They needed more communication to reach a compromise. Why can the government not manage the credibility gap between vendors and communities?

An Environmental Impact Assessment (EIA) has been stipulated by the EPA, Executive Yuan (Taiwan's executive branch of government). However, why can the Taiwan government not effectively manage negative perceptions of the EIA process? The main reason is political considerations. The former president of Taiwan, Lee Teng-Hui, believed Taiwan, as a democratic country, should empower the communities with the right of "civil disobedience." Any person in Taiwan society should have the right to actively refuse to obey certain regulations, decisions, laws, and commands by government. The primary tactics of civil disobedience should be nonviolent and peaceful resistance. He ordered government officials to adopt a peaceful and impartial way to deal with the credibility discrepancy between vendors and communities. Therefore, the only response by policemen who are charged with the order of public protest against vendors is to define the boundary of demonstrations and protest, and protect the right of civil resistance to public construction.

Table 9. Survey Results on the Credibility Gap

	Definitely do not trust	Do not trust	Trust	Definitely trust	Unknown
1. The power plant site selection: Did the vendor truly want to improve the local community's prosperity?					
Power stations	13.1% (34)	30.8% (80)	39.6% (103)	3.5% (9)	34% (13.1)
Incinerators	22.7% (174)	33.2% (254)	25% (191)	4.6% (35)	14.5% (111)
2. The site selection: Did the vendor truly want to resolve the pollution and safety issues?					
Power stations	10.8% (28)	26.5% (69)	40.8% (106)	4.6% (12)	17.3% (45)
Incinerators	13.6% (104)	20.5% (157)	40.7% (311)	6.8% (52)	18.4% (141)

- *Problem of Local Factions*

In local Taiwanese politics, there were two problems arising in NIMBY confrontation. One problem was the underground activities emerging during the public facility construction because of the LULU's willingness to provide monetary compensation for environmental effects. These underground groups profited by helping the vendor settle the problem. As a result, their involvement tended towards illegal means of resolving the conflict. From our interviews, we found a couple of illicit occurrences but they were not serious. According to the survey report, 30 percent of the participants had experienced or heard of this phenomenon. Of the participants, 55 percent thought politicians had a hand in the outcomes. These results suggest the underground situation today is much better than in the past.

The other problem was the news coverage and local politics during the construction process. Local politicians need voters to win the election for the next term of office. Public protest and demonstrations always become "headlines" of newspapers, radios, and televisions. Thus, local politicians usually tried to capitalize on the opportunity to build his or her popularity with the local community and participated in the confrontation to attract voters during news reporting. According to our study, in almost all of the power plant's surrounding areas, local politicians took part. In the name of trying to peacefully strike a compromise between the vendor and the local community, these politicians actually led the confrontation, politicising the NIMBY syndrome. Especially during the election period, these local politicians used the situation to build their momentum in the community. Instead of explaining reasons for site selection and educating residents how to deal with the negative risks of NIMBY facilities, local politicians preferred to ingratiate themselves with the local interests and thus tend towards populism.

To win voters' support and maintain the majority dominance in the Legislative Yuan, the former ruling party, Kuomintang, asked responsible government officials to adopt "soft" and pleasing means, including a compensation strategy, to cope with the protesting citizens. The result of this populism and local factionalism leads to a distorted expression: "The more candy you extort from government, the more money you get." Therefore, negotiated compensation measures turned out to be a major strategy to solve the intense conflict between the facilities vendors and the communities through local politicians.

BREAKING THROUGH THE GRIDLOCK OF THE LULU SYNDROME

Combining the above research results, we conclude NIMBY syndrome existed in the field of Taiwan's LULU site selection and construction. Compared to modern American society, the facts here are more complicated. In sum, it took two types of factors, economic and noneconomic, to explain the phenomenon, which is more than usual (Dear, 1992; Mazmanian & Morell, 1994; Rabe, 1994; Takahashi & Dear, 1997). In general, considering the privatization of power plants and solid waste incinerators in Taiwan society, the main factor causing NIMBY syndrome was the host community's concerns about the LULU site selection. It was perceived to harm the community in not only health consequences and property values, but also in their living standards. In this case study, we noted it was difficult to strike a fair environmental compensation.

We also noted a lack of host-community participation in policy making did not foster society's trust in the process. To complicate this, underground elements and local politicians became involved to capitalize on the chaos. In a densely populated, narrow island such as Taiwan, this situation was further taxed by the fact that the economy was highly developed; and, subsequently, the people's standards were higher than the developing country's standards. In this scenario, it was therefore necessary to use a different policy approach to devise a settlement. What is the best way to settle the LULU syndrome? Dear (1992) suggested "community-based, government-based and court-based strategies." Rabe (1994) indicated in hazardous waste site selection in Canada's and the United States's facility-site selection approach, usually there are two strategies: the regulatory approach and the market approach. Because two types of factors, economic or noneconomic, influence the success of LULU sites, we offer three approaches: (1) government-regulatory, (2) market-based, and (3) community-governance strategies.

(1) Government regulatory strategies: the government usually forcibly takes action to select the site. Both the vendor and host community must obey the government's determination. If a violation occurs, there is administrative law to punish offenders or arbitration in court for settlement. This is what Rabe (1994) called the "coercive siting process." This is also Taiwan's government's present adopted model for settling the NIMBY syndrome. Because people do not welcome the facility, the government has to intervene to force the vendor or facility builder to pay more attention to the effects on health, property, and living standards of the local community in the hope of minimizing the negative reaction. Thus, (A) the strategy for resolving health and property risks is as follows: (i) the government can force the power plant to

periodically sponsor free health check-ups for the host community; (ii) the government can monitor the land values and use near the high-voltage tower and confiscate it if necessary to see that people are fairly compensated for avoiding any further monetary conflict. (B) The strategy for resolving living standard effects: the relevant government agency has to establish the air, water, and waste-material pollution control as recourse for avoiding the occurrence of accidental incidents and improving the security system. The government's labour security–management department must periodically inspect the gas pipe and liquid natural gas security. The government also has to sponsor an association between the community and the power plant and incinerator vendors to improve relations and improve the community's landscape, making the unacceptable facility a welcome one for the host community.

(2) Market-based strategy: in the United States, many state governments require the toxic-waste processing facility to pay NIMBY taxes (Levinson, 1999). It is currently impossible to pass the law of NIMBY taxes in the legislative body in Taiwan.[3] Actually, a "gentlemen's agreement," an informal and mutual agreement between representatives of the host community and vendors of LULU facilities based on volunteering and negotiation, is encouraged by the Environmental Protection Administration under Taiwan's Executive Yuan. Inhaber's (1998) auction strategies and Mazmanian and Morell's (1994) site-selection contract approach are similar. They all emphasize self-determination and voluntary action by the host community. It is best to allow the host community to fully participate in the process of site selection, and to let the people understand all the detailed information about the construction and the business management. Depending on what is agreed on by all parties, the contractor then should take responsibility for the environmental pollution collection or avoid the responsibilities of the destruction of the living standards, safety, and compensation.

---

[3] The reasons are twofold: First, too many types of NIMBY facilities do not allow a plausible tax rate and tax reduction formula. For instance, how do we calculate the negative impacts of NIMBY facilities, such as incinerators, power stations, those for transportation, recreation, water supply, social service, etc., on residents? Different facilities pose different threats to people. Second, NIMBY taxing the law-making process is tedious in the Legislative Yuan. Diverse interest groups and legislators usually play "dirty" bargaining games, or the so-called pork-barrel legislation in the American legislative process. So the EPA prefers to delegate authority to the local government to make decisions on compensating neighbouring communities on a case-by-case basis instead of a general taxing clause.

(3) Community participatory strategies: Due to the multiple factors involved in LULU facilities, settling NIMBY syndrome in Taiwan is more complicated than most developing countries would encounter and requires a more organized strategy. Thus, the community-governance mode is suggested to resolve the dilemma of NIMBY conflicts. How is it possible to manage factionalism and local faction leaders around site-selection processes through the community-governance approach? We should understand there is a dramatic difference between the community governance mode and local factionalism. Parochial local faction leaders are only concerned with self-interest through the "black box" of the NIMBY conflict-management process. However, the community-governance approach, under the jurisdiction of the Community Empowerment Commission under the Executive Yuan, emphasized the roles of all the residents participating in an open and transparent decision-making process based on public interest and civil responsibility. The community-empowerment movement has been incorporated into national development policy,[4] which is supervised by the Executive Yuan. Many communities in this island country have accumulated experience of operating comanagement and of the community-governance mode for roughly two decades. Of course, it is necessary to heighten citizen participation to strengthen the function of the community-governance mode.

However, Taiwanese believe the society needs a community governance–oriented system to include persons with community-empowerment talents. The public-facility site selection, under the perspective of the community-governance approach, has to increase the host community's participation to minimize NIMBY syndrome and increase mutual trust, avoiding physical forms of confrontation, such as that in development in Chula Vista, California, south of San Diego. Young (1990, pp. 25–26) proposed risk communication should be the primary means to settle the NIMBY syndrome.

Paehlke & Torgerson (1990) analyzed the relationship between toxic waste and the administrative state, and concluded NIMBY syndrome should be handled by participatory management. Therefore, in Taiwan, we suggest adopting the following steps for the process. (1) The strategy to resolve the problem of the expert dictatorship policy: first, we have to assess the effects on the environment, organize a hearing, and make it a legal process. Second, we have to realize the executive legal procedure about the hearing and broadly disseminate and disclose the information to the public. Third, we have to

---

[4] According to my estimation, over 3,000 communities are included in the "Six-Star" Formosa Community Program, August 31, 2008, sixstar.cca.gov.tw/

establish the referendum guidelines, allowing the host community to have the right to vote in the referendum. (2) The strategy to settle the credibility gap: the government procedure has to settle the power-plant vendor's neighbourhood dispute publicly and openly and carry it out in a legal manner. Implementing this strategy relies on clear and effective communication between the government, the vendor, and the host community. (3) The strategy to solve the problem of environmental compensation: this market-based and voluntary approach is worth adopting, especially in the compensation for the effects on the environment. Since the government cannot form a law to request that the vendor or developer offer a monetary compensation, it can only set up a fair and just system to guide and form a strict judgment on the effects on the environment. Once settled, the government has to fully support the decision. It does not have to consider the compensation problem unless residents produce results from a doctor's physical examination and show they have really suffered from the facility. Therefore, the government can encourage a compromise between the vendor and the people, and provide suitable compensation for the people.

The government has to help the vendor and the neighbourhood systematise the environment's compensation system and set up a tripartite committee, consisting of a community representative, the vendor, and a moderator (Stallen, 1991, pp. 59–60) to promote cooperation and monitor the pollution. This will establish the basis or improve the living environment and the welfare of the host community. Meanwhile, of the two forms of compensation—monetary or nonmonetary—we recommend not choosing the monetary option because this would create a situation where people come to expect compensation all the time, which over time becomes problematic because it lures dishonest elements. The vendor should create job opportunities for the community and beautify the environment. It should establish the welfare facility and take care of the elderly and disabled or low-income families or set up a hospital or clinic.

The foregoing strategy basically could settle six factors, regarding three economic factors, i.e., the health and property risks, the risk of destroying the quality of life, the compensation for any effects on the environment; meanwhile three noneconomic factors, i.e., avoiding the tyranny of the expert in policymaking, and narrowing the credibility gap. In Taiwan, however, the local politicisation problem, which in site selection in Taiwan tends to be the hardest factor in the NIMBY syndrome to settle. The local LULU officials that lead the community into confrontation with the central government's public facility's policy usually act as the NIMBY group's leader, forgetting their duty is to execute the policy. These local leaders should obey the central government's directives but instead, choose to lead the people into serious confrontations to show they "stand with the people." The reason is obvious. Because the people elect them, they are obligated to represent them. If they seek reelection, they also are under pressure to realise the promises they made to their constituents. Therefore, this becomes a phenomenon of populism. This is the toughest challenge for Taiwan's government policy makers and urban planners.

REFERENCES

Brion, D. J. (1991). *Essential industry and the NIMBY phenomenon*. New York: Quorum Books.

Bryant, B. I., & Mohai, P. (Eds.). (1992). *Race and the incidence of environmental hazards: A time for discourse*. Boulder, CO: Westview.

Chiou, C.-T. 丘昌泰 (2002). Cong linbi qingjie dao yingbi xiaoying: Taiwan huanbao kangzheng de wenti yu chulu 從鄰避情結到迎臂效應：台灣環保抗爭的問題與出路 From (NIMBY to YIMBY: Problems and solutions of environmental movement in Taiwan). *Zhengzhi kexue luncong* (Political science review), *17*, 33–56.

Connor, D. M. (1988). Breaking through the NIMBY syndrome. *Civil Engineering, 58*(12), 69–71.

Curic, T. T., & Bunting, T. E. (2006). Does compatible mean same as? Lessons learned from the residential intensification of surplus hydro lands in four older suburban neighborhoods in the city of Toronto. *Canadian Journal of Urban Research, 15*(2), 202–224.

Dear, M. (1992). Understanding and overcoming the NIMBY syndrome. *Journal of the American Planning Association, 58*(3), 288–300.

Dorshimer, K. R. (1996). Siting major projects & the NIMBY phenomenon: The Decker energy project in Charlotte, Michigan. *Economic Development Review, 14*(1), 60–63.

Feinerman, E., Finkelshtain, I., & Kan, I. (2004). On a political solution to the NIMBY conflict. *The American Economic Review, 94*(1), 369–381.

Gates, C. (1999). Community governance. *Futures, 31*, 519–525.

Gould, K. A. (2002). Review essay: The democratization of environmental policy. *Rural Sociology, 67*(1), 122–139.

Hall, R. (1980). *Great planning disasters*. Berkeley, CA: University of California Press.

Halstead, J. M., Luloff, A. E., & Myers, S. D. (1993). An examination of the NIMBY syndrome: Why not in my backyard? *Journal of the Community Development Society, 24*(1), 88–102.

Hoyman, M., & Weinberg, M. (2006). The process of policy innovation: Prison sitings in rural North Carolina. *Policy Studies Journal, 34*(1), 95–112.

Hunter, S., & Leyden, K. M. (1995). Beyond NIMBY: Explaining opposition to hazardous waste facilities. *Policy Studies Journal, 23*(4), 601–619.

Inhaber, H. (1992). Of LULUs, NIMBYs, and NIMTOOs. *Public Interest, 107*, 52–65.

Inhaber, H. (1998). *Slaying the N.I.M.B.Y. dragon*. New Brunswick, NJ: Transaction Publishers.

Kim, D. S. (2000). Another look at the NIMBY phenomenon. *Health & Social Work, 25*(2), 146–148.

Korten, D. C. (1995). *When corporations rule the world*. West Hartford, CT: Kumarian Press.

Kurland, O. M. (1992). Risk communication, mitigation and uncertainty. *Risk Management, 39*(12).

Lake, R. W. (1987). *Resolving location conflicts.* New Brunswick, NJ: Rutgers University Press.

Lake, R. W. (1993). Rethinking NIMBY. *Journal of the American Planning Association, 59*(1), 87–93.

Lee, Y.-Z. 李永展. (1997). Linbi zhenghouqun zhi jiexi 鄰避徵候群之解析 (An analysis of NIMBY syndrome). *Dushi yu Jihua* 都市與計畫 (Urban and Planning), *24*(1), 69–79 (in Chinese).

Lehr, J., & Inhaber, H. (2003). Comment on: A creative solution to the NIMBY problem/reply. *Ground Water, 41*(6).

Lesbirel, S. H. (2000). *NIMBY politics in Japan.* Ithaca, NY: Cornell University Press.

Levinson, A. (1999). NIMBY taxes matter: The case of state hazardous waste disposal taxes. *Journal of Public Economics, 74*, 31–51.

Mazmanian, D. A., & Morell, D. (1994). The NIMBY syndrome: Facility siting and the failure of democratic discourse. In N. J. Vig & M. E. Kraft (Eds.), *Environmental policy in the 1990s: Toward a new agenda.* Washington, DC: CQ Press.

McAvoy, G. E. (1998). Partisan probing and democratic decision-making: Rethinking the NIMBY syndrome. *Policy Studies Journal, 26*(2), 274–292.

McAvoy, G. E. (1999). *Controlling technology: Citizen rationality and the NIMBY syndrome.* Washington, DC: Georgetown University Press.

Neuzil, C. E. (2003). Comment on: A creative solution to the NIMBY problem. *Ground Water, 41*(6).

O'Hare, M. B., & Sanderson, D. (1988). *Facility siting and public opposition.* New York: Nostrand Reinhold Company.

Paehlke, R., & Torgerson, D. (1990). Toxic waste and the administrative state: NIMBY syndrome or participatory management. In R. Paehlke & D. Torgerson (Eds.), *Managing Leviathan: Environmental politics and the administrative state.* London: Belhaven Press.

Popper, D. (1987). The environmentalist and the LULU. In R. Lake (Ed.), *Resolving locational conflict.* New Brunswick, NJ: Center for Urban Policy Research.

Putnam, R. D. (1993). *Making democracy work: Civic traditions in modern Italy.* Princeton, NJ: Princeton University Press.

Putnam, R. D. (2000). *Bowling alone: The collapse and revival of American community.* New York: Simon & Schuster.

Rabe, B. G. (1994). *Beyond NIMBY: Hazardous waste siting in Canada and the United States.* Washington, DC: The Brookings Institution.

Richman, B. D., & Boerner, C. (2006). A transaction cost economizing approach to regulation: Understanding the NIMBY problem and improving regulatory responses. *Yale Journal on Regulation, 23*(1), 30–76.

Sechrest, D. K. (1992). Locating prisons: Open versus closed approaches to siting. *Crime and Delinquency, 38*, 88–104.

Sellers, M. P. (1993). NIMBY: A case study in conflict politics. *Public Administration Quarterly, 16*(4), 460–477.

Sénécal, G., & Reyburn, S. (2006). The NIMBY syndrome and the health of communities. *Canadian Journal of Urban Research, 15*(2), 244–263.

Stallen, P. J. M. (1991). Developing communications about risks of major industrial accidents in the Netherlands. In R. E. Kasperson & P. J. M. Stallen (Eds.), *Communicating risks to the public*. Boston: Kluwer Academic Publishers.

Takahashi, L. M. (1997). The socio-spatial stigmatization of homelessness and HIV/AIDS: Toward an explanation of the NIMBY syndrome. *Social Science Medicine, 45*(6), 903–914.

Takahashi, L. M., & Dear, M. J. (1997). The changing dynamics of community opposition to human service facilities. *Journal of the American Planning Association, 63*(1), 79–93.

Vittes, M. E., Pollock, P. H., III, & Lilie, S. A. (1993). Factors contributing to NIMBY attitudes. *Waste Management, 13*, 125–129.

Waugh, W. L., Jr. (2002). Valuing public participation in policy making. *Public Administration Review, 62*(3), 379–382.

Weisberg, B. (1993). One city's approach to NIMBY. *Journal of the American Planning Association, 93*(1), 93–97.

Yarzebinski, J. A. (1992). Handling the "not in my backyard" syndrome: A role for the economic developer. *Economic Development Review, 10*(3), 35–40.

Young, S. (1990). Combating NIMBY with risk communication. *Public Relations Quarterly, 35*(2), 22–26.

Zippay, A. L. (2007). Psychiatric residences: Notification, NIMBY, and neighborhood relations. *Psychiatric Services, 58*(1), 109–114.

Challenges of Managing NIMBYism in Hong Kong

Kin-che Lam, Wai-ying Lee, Tung Fung, and Lai-yan Woo

INTRODUCTION

The siting of Locally Unwanted Land Uses (LULUs) is one of the critical issues confounding Hong Kong's long-term development. To cater to the territory's growing population and increasing affluence, Hong Kong has seen a demand for more power generation and waste-disposal facilities, as well as correctional institutes and medical centres to treat infectious diseases. The problem of siting these LULUs is probably as acute, if not more so, in Hong Kong as in other metropolitan areas because of the severe planning constraints imposed by the territory's small size and the increasing environmental awareness of the general public.

This paper reports the initial findings of a public-policy research project, supported by the Hong Kong Research Council, on NIMBYism in Hong Kong. The objectives of the current paper are to:

(a) elucidate how NIMBYism has arisen in the unique political, social, economic, and geographical context of Hong Kong;
(b) explore how conflicts arising from LULUs might be resolved.

This research began with a conceptualisation of the Not in My BackYard* (NIMBY) phenomenon in various parts of the world and in Hong Kong (Lai et al., 2007). A two-stage questionnaire survey was undertaken to gauge the

* The research reported in this paper was supported by a research grant provided by the Hong Kong Research Grants Council (4008-PPR20051). We also wish to thank Ms. Mary Felley for her invaluable assistance in proofreading this paper.

public's understanding of the NIMBY phenomenon and their perception of the need for, and fear of, LULUs in Hong Kong and in their own neighbourhood. The first was a telephone survey undertaken in early May 2007 in which a total of 1,002 interviews were successfully completed. The respondents were randomly selected from all geographical districts of Hong Kong to reflect the views of the population of Hong Kong at large. Subsequent to the first territory-wide survey, a similar questionnaire survey was undertaken in Tuen Mun, an area of Hong Kong with a disproportionate share of LULUs. In this survey, 752 residents were successfully interviewed on the streets in parts of Tuen Mun about their perception of the LULUs in their neighbourhood and how the conflicts might be resolved. Owing to resource constraints, the questions posed in both surveys focused on environmental LULUs; however, the researchers did not exclude any answers pointing to LULUs of other types in Hong Kong.

THE LOCAL CONTEXT

While NIMBY is a worldwide phenomenon, the mode of its emergence, the dynamics between the key players and the means of resolution are shaped by the local geographical, political, and socioeconomic context.

Hong Kong is a special administrative region (SAR) of China, essentially a small city-state that enjoys a high level of autonomy in all areas of public administration except defence and foreign policy. Hong Kong is run by the SAR government, which is largely "administration-led" with little power bestowed to the local district councils in planning matters (Leverett et al., 2007a). In the legislative council, the political parties are reactive rather than proactive when it comes to Hong Kong's environmental policy agenda. Infrastructure building in Hong Kong is mostly initiated by the government, which is vested with the authority to plan in consultation with statutory and nonstatutory boards and consultative committees and local district councils. These boards and consultative committees are largely composed of non-official members of the public appointed by the government. The main function of district councils is to advise the government on matters affecting the well-being of the people and on the provision and use of public facilities and services. While members of the district council can offer views on proposed developments, the final decision rests with the government. This institutional setup has been criticised by some as being too centralised and top down (Leverett et al., 2007a), as a result of which local residents may feel alienated from central policy and plan making and seldom gain a sense of control of their immediate environment (Ng, 2004).

The difficulty of siting LULUs in Hong Kong is aggravated by the physical terrain and small size of the city. With only 1,104 km^2 of land, Hong Kong is home to 6.9 million people. Hong Kong's hilly terrain forces urban development to be concentrated on about 22 percent of the total land area. The areas outside the major urban developments are mostly too hilly and hence costly to develop, and most are designated country parks and water-gathering grounds where major developments of any type, let alone LULUs, are strictly regulated. The physical terrain, the valley pockets and sea inlets in some parts of Hong Kong are not favourable for the dispersion of air and water pollutants.

Due to the easterly prevailing winds and differences in topography, not all of Hong Kong's eighteen electoral districts have similar environmental capacities. Given the prevailing easterly winds, most major air-polluting sources are located in the western part of Hong Kong. Topographical and water-circulation considerations have also excluded certain districts, such as Shatin, as potential sites for major air- and water-discharge facilities. Hence, it would probably be undesirable, at least from an environmental perspective, to equally distribute environmental LULUs across the 18 districts of Hong Kong.

Hong Kong is the world's 11th largest trading economy, with Mainland China being its most significant trading partner. While Hong Kong has enjoyed some significant economic growth in recent years, there are evidences of a widening social gap between the rich and the poor (CSD, 2007). This has nurtured a sense of discontent with the government. Combined with the concentration of LULUs in some districts, social segregation has resulted in a labelling effect of the community and has nurtured grievances, mistrust, and a sense of injustice.

At the same time, there is, among the general public, an increasing demand for environmental quality and democratisation of the political system. Local civil societies are becoming more vocal in their demand for quality of life, open and transparent government, and greater social justice.

While the Hong Kong SAR government is seen by many as efficient and free from corruption and malpractices, the current planning approach and environmental-assessment practices are apparently inadequate to ease tensions and conflicts arising from LULU proposals. To some (Ng, 2004), the planning approach is too technocratic and rational, top-down, and insensitive to local needs. Likewise, the environmental impact assessment (EIA) process, which is effective in preempting pollution problems (Lam & Brown, 1997) and is becoming increasingly transparent and accountable (Leverett et al., 2007b), has nonetheless been criticised by some as being too focused on environmental technicalities and not adequately responsive to local concerns. There are also criticisms (Leverett et al., 2007b) that some EIAs fail to give sufficient justification for the choice of a particular site or route and that consultation is conducted at a time when project planning has gained so much momentum that the project can hardly be reversed. Such a planning approach has merits in terms of the optimality of the site from the environmental perspective, social and political considerations are, however, often ignored. Once the "optimal site" is identified, there is little room for change or negotiation (Kuhn & Ballard, 1998). In other countries, LULU siting has gradually evolved from the top-down approach to an increasingly decentralised and participatory approach (Lesbirel & Shaw, 2005).

Unlike the government, which sees the siting of LULUs as a technocratic issue, members of the public have their own beliefs, priorities, and sense of fairness that shape their level of trust in the government. Furthermore, the government has no clearly defined strategy to manage siting conflicts. All these have created a scene of spatial inequality, nurtured seeds of discontent, and crafted a milieu that makes it difficult to resolve conflicts arising from the siting of LULUs.

NIMBYISM AS REVEALED IN THE TERRITORY-WIDE SURVEY

As expected, findings of the territory-wide telephone survey confirmed that the phenomenon of NIMBYism is commonly found among the general public in Hong Kong. The term "NIMBY" was initially coined by Popper (1981) to refer to any LULU that may be regionally or nationally needed but is considered objectionable to the people who live nearby. This was further expounded by Wolsink (1994) who observed that "everyone acknowledges the importance of the public good, but not everyone is prepared to make a personal contribution, in this case by co-operating in the construction of an installation in one's neighborhood." This characterisation is borne out by the territory-wide survey, which showed that significantly more people envisage the need for having a particular LULU for the whole of Hong Kong than for their region or local district (Figure 1). The perceived need and public acceptance are also markedly greater for facilities that most people use (e.g., garbage stations) than those that people do not readily associate with, such as the chemical waste-treatment facility and waste incinerator (CEPRM, 2007a).

Our survey (CEPRM, 2007a) also showed that slightly more people opine that it is fair to site LULUs according to the need of different districts or to evenly distribute them in space. While the planning approach of spatially dispersing LULUs is a potential candidate for overcoming local resistance (Popper, 1981) and has been followed to a small extent in countries such as

Figure 1. Public Perceived Need for Locally Unwanted Facilities in Hong Kong, the Region and Local District

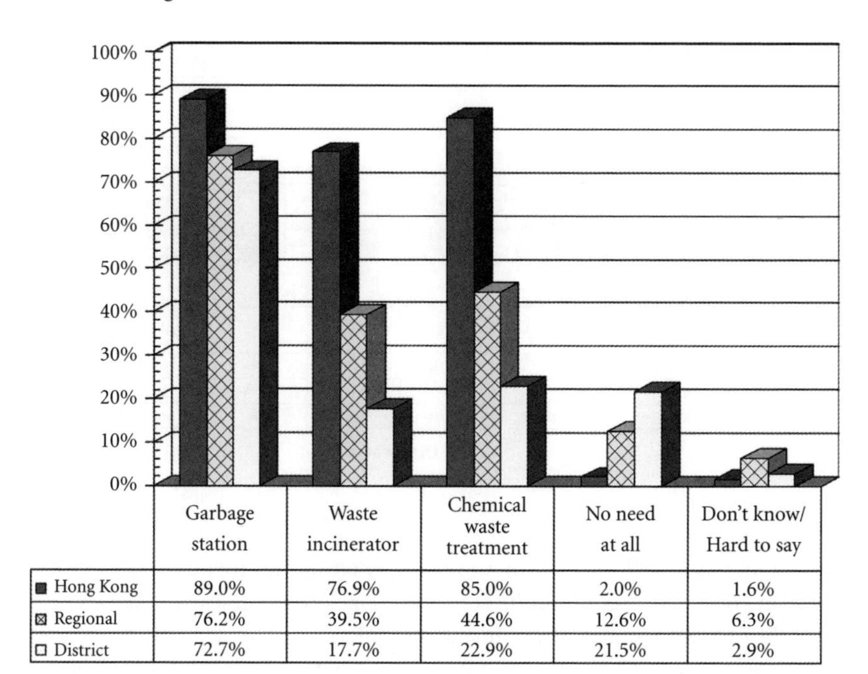

	Garbage station	Waste incinerator	Chemical waste treatment	No need at all	Don't know/ Hard to say
Hong Kong	89.0%	76.9%	85.0%	2.0%	1.6%
Regional	76.2%	39.5%	44.6%	12.6%	6.3%
District	72.7%	17.7%	22.9%	21.5%	2.9%

Japan (Macintyre & Tashiro, 2000), its implementation in Hong Kong is not likely to be viable because of the uneven environmental assimilative capacity between districts. Furthermore, some districts in Hong Kong are not large enough to generate sufficient demand to justify a facility to meet their own needs.

The challenge in siting LULUs is further confounded in Hong Kong by the lack of trust in the government. Generally speaking, the public has more trust in civil societies and environmental nongovernmental organizations (ENGOs) than in the government and the private sector (Figure 2). While public participation is often seen as a way to foster trust (Baxter et al., 1999), our survey shows that only a small percentage (13.3 percent) of the respondents agreed and strongly supported the statement that the consultation undertaken by the government is adequate. The lack of trust in the government in managing LULU facilities may have significant implications. The literature abounds with cases in which social distrust increases the perceived risk of a facility (Groothuis & Miller, 1997) and leads to strong opposition to the proposal (Yoo, 1996).

Figure 2. The Degree of Trust That the General Public in Hong Kong Places on the Major Actors Involved in LULU Siting

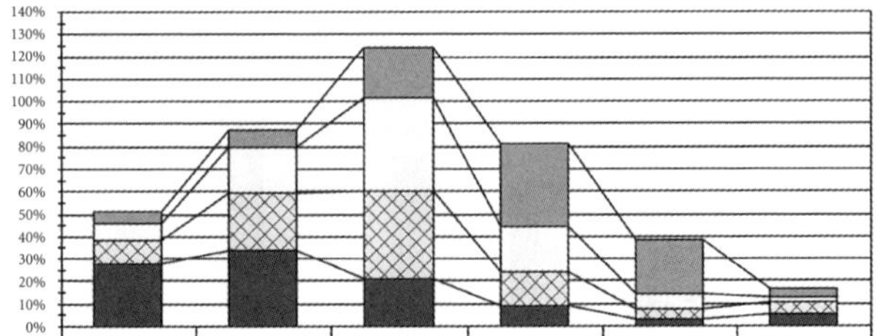

	1 (No risk at all)	2	3	4	5 (Trust a lot)	Don't know/ Hard to say
Community groups (including green groups)	4.9%	7.2%	22.6%	37.1%	24.4%	3.7%
Government	7.7%	20.9%	41.6%	20.2%	6.9%	2.6%
Public-private organisations	10.3%	25.4%	39.4%	14.7%	4.5%	5.4%
Private companies	28.1%	33.8%	20.8%	9.3%	2.8%	5.0%

Degree of trust

LULUs are well known for imposing externalities involuntarily on the residents of the local community (Quah & Tan, 2002). These externalities may include environmental impacts, health and safety impacts, social impacts, and economic impacts. Our survey results indicate that more people are concerned with environmental and health impacts and associated risks than with social and economic losses (CEPRM, 2007a). The relative importance of externalities as perceived by the Hong Kong public is similar to those observed in Japan with respect to waste-management facilities (Rahardyan et al., 2004). However, the perceived externalities in Taiwan (Chiou, 2005) are somewhat different: in descending order, they are declining property value, negative health impacts, and reduced crop productivity for incinerators and power stations. The rank order of these public concerns may have implications for the mitigation measures that can be adopted to alleviate public opposition.

FINDINGS OF THE TUEN MUN SURVEY

A single district in Hong Kong of a total of 18, Tuen Mun is home to many of Hong Kong's LULUs. It has one of the territory's two power stations, one of its two major psychiatric hospitals, and one of its three strategic landfills. Tuen Mun was also the site of one of the three major refugee camps in the 1990s. Located within the boundary of Tuen Mun are also Hong Kong's only aviation fuel–receiving facility, steel plant, river-trade terminal, and a large waste-recycling park. There are also plans to build a mega columbarium-cum-crematorium and a sewage-sludge incinerator that would serve the rest of Hong Kong.

Located on the western extremity of Hong Kong's mainland section, Tuen Mun is seen as the most appropriate site for major air pollution sources because it is on the downwind side of the prevailing wind. It is also close to Hong Kong's international airport and the Pearl River Estuary, making it a favoured site for handling aviation fuel and river-cargo traffic. In addition to negative physical impacts, this disproportionate share of LULUs in Tuen Mun generates other negative externalities such as loss in aesthetic values and decline in community attractiveness.

Probably because of financial stringency at the time of development, the community facilities of Tuen Mun such as town halls, libraries, urban parks, sports grounds, and the light-rail system are not as well furnished and impressive as those in other new towns of Hong Kong. It was not until the early 2000s that Tuen Mun was linked up with the rest of Kowloon by a mass-transit railway system. For various reasons, Tuen Mun has a socioeconomic profile that is not on par with the Hong Kong average. For example, according to the 2006 bi-census, Tuen Mun has a smaller percentage of people who have received secondary education (15.8 percent vs. 21.9 percent) and a lower household median income ($15,000 vs. $17,500 per month) than Hong Kong as a whole. These factors only reinforce the negative image and stigma of Tuen Mun.

Despite the concentration of LULUs in Tuen Mun, it is paradoxical to note from the questionnaire survey (CEPRM, 2007b) that 55 percent of Tuen Mun residents were not aware of the occurrence of LULUs in their district unless prompted by the interviewer. This is reinforced by the survey finding that 47 percent of the respondents do not agree with the commonly held belief that Tuen Mun has a disproportionate share of LULUs compared to other districts. Furthermore, only 14 percent of the Tuen Mun respondents could correctly name a LULU in their neighbourhood.

There is no apparent single reason to account for the observed paradox. The phenomenon can be ascribed to a number of possible factors. Firstly, Tuen Mun residents are probably more concerned with making a living and increasing household income than with the neighbourhood environment. Secondly, LULUs are not evenly dispersed within Tuen Mun; most of them are concentrated in a special industrial zone, known as Area 38, situated purposely on the outskirts of Tuen Mun Township. This special industrial zone is about 1.8 kilometers from the town centre and one kilometer from the nearest large housing development.

Figure 3. The Degree of Trust of Tuen Mun Residents in Nongovernmental Organizations, Legislative and District Councils, Government, Public-Private Organizations, and Private Companies

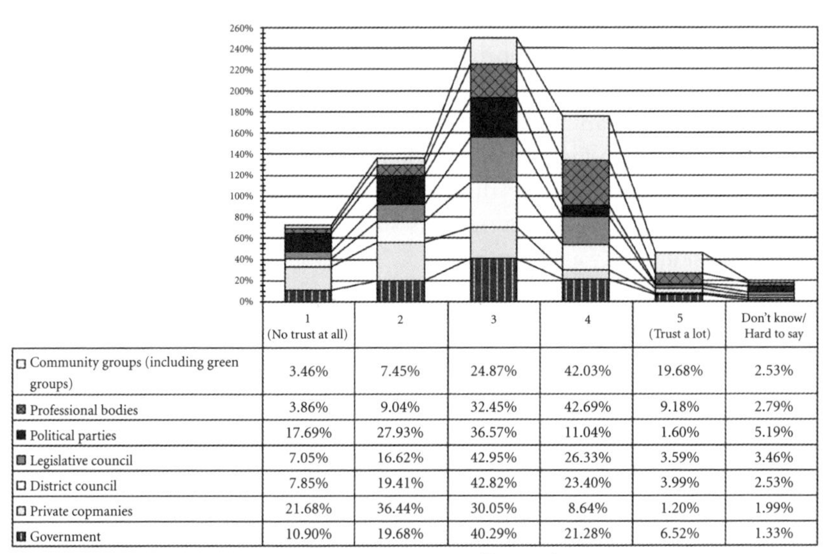

	1 (No trust at all)	2	3	4	5 (Trust a lot)	Don't know/ Hard to say
Community groups (including green groups)	3.46%	7.45%	24.87%	42.03%	19.68%	2.53%
Professional bodies	3.86%	9.04%	32.45%	42.69%	9.18%	2.79%
Political parties	17.69%	27.93%	36.57%	11.04%	1.60%	5.19%
Legislative council	7.05%	16.62%	42.95%	26.33%	3.59%	3.46%
District council	7.85%	19.41%	42.82%	23.40%	3.99%	2.53%
Private copmanies	21.68%	36.44%	30.05%	8.64%	1.20%	1.99%
Government	10.90%	19.68%	40.29%	21.28%	6.52%	1.33%

Degree of trust

The lack of awareness of LULUs among Tuen Mun residents is paralleled by their lack of knowledge of the planning process involving LULUs. The Tuen Mun survey shows that 86 percent of the respondents did not know how LULUs are planned and sited, and 31 percent did not have trust or a lot of trust in the government (Figure 3). It merits mentioning from the same figure that the level of trust in the local district council, where LULU proposals are discussed, is in fact low and very close to that of the government.

Probably because of the lack of trust in the government and the establishment, over 58 percent of the respondents cited the media, such as newspapers and TV, as their main source of information about LULUs. This is in stark contrast to the mere two percent of respondents who reported knowing LULUs through town hall meetings organized by the government. The overreliance on mass media as the source of information probably explains why some local residents named LULUs in Tuen Mun that had been widely reported in the news media but were not yet in existence at the time of the survey.

When Tuen Mun residents were asked whether they would welcome some specific types of LULUs in their district, the majority of them opposed having more LULUs despite the fact that some were aware of the benefits those facilities could bring to society. In fact, their feelings were so negative that they opposed LULUs of all types (CEPRM, 2007b). This negative attitude is probably linked to the finding that 79 percent and 68 percent, respectively, of respondents reported that consultation undertaken by the government was inadequate or ineffective. Another 27 percent reported that they had no faith in the local district council in handling matters related to LULUs.

It is nonetheless interesting to note that what was in doubt is not the government's technical competence, but the risks and uncertainty associated with hazardous installations (Figure 4 and Table 1). Similar to what is commonly reported in the literature, LULUs that may carry risks of low frequency and catastrophic consequences are those that elicit the greatest fear and opposition (Popper, 1987; Slovic, 1987). This finding lends support to the findings of previous studies that the public reacts differently to different types of LULUs, and that those that impose high impacts and risks to humans are the least welcome.

POLICY IMPLICATIONS

The findings of the two surveys, undertaken in the whole of Hong Kong and in a district seen by many as having a disproportionate share of LULUs, suggest that NIMBYism is an issue and a challenge to infrastructure building. The results also suggest a level of mistrust and breakdown in communications between the planning authority and the host community. Whilst members of the public do not doubt the technical competence of the government, there is a notable absence of trust in the government and in the mechanisms of communication and consultation. In the absence of trust, the majority of the public rely on the news media as the source of information.

The research findings confirm that different types of LULUs invoke different levels of fear and should be treated differently. Generally speaking, the public is less resistant to LULUs that create environmental nuisances than those that bring about uncertain and uncontrollable risks (Slovic, 1987), such as hazardous installations and infectious disease medical centres.

The survey results call into question the efficacy of the current planning process and public consultation strategy in informing and engaging the public. It is well known that the news media play an important role in risk communication (Johnson-Cartee et al., 1992/1993), and if and when the risks are misunderstood and amplified, mistrust can be accentuated (Quah & Tan, 2002; Upreti & van der Horst, 2004). The lack of trust may result in impasses that are extremely difficult to resolve.

Figure 4. Risk Level of Different Locally Unwanted Facilities as Perceived by Tuen Mun Residents

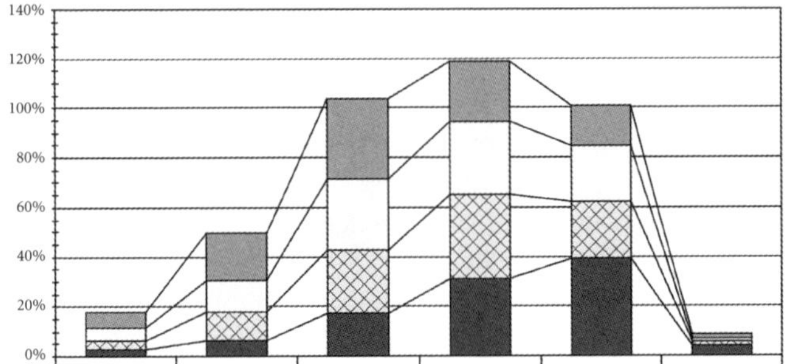

	1 (No risk at all)	2	3	4	5 (Very risky)	Don't know/ Hard to say
Landfill	6.78%	18.88%	32.45%	23.94%	16.22%	1.73%
Incinerator	4.79%	12.90%	28.99%	29.52%	22.47%	1.33%
Coal-fired power plant	3.72%	11.44%	25.53%	34.31%	23.27%	1.73%
Aviation fuel–receiving facilities	2.79%	6.38%	17.02%	30.85%	39.10%	3.86%

Degree of perceived risk associated with different types of LULUs

Table 1. Tuen Mun Residents' Perception of Risk Characteristics

Statement	Don't agree/ Totally don't agree	Somewhat agree	Agree/ Totally agree	Don't know/ Hard to say	Decline to respond
The facility will bring catastrophic effects if accidents occur	5.05%	9.31%	84.85%	0.80%	0.00%
Environmental impacts arising from the facility are not easy to reduce	4.12%	11.30%	83.11%	1.46%	0.00%
The risks associated with the facility would fill people with fear and dread	10.11%	17.69%	71.54%	0.66%	0.00%
The technology of the facility may not be reliable	30.45%	27.66%	35.50%	6.38%	0.00%
Public are not familiar with the impacts and risks associated with the facility	9.71%	21.81%	65.02%	3.46%	0.00%
The facility may impose impacts and risks upon future generations	9.04%	12.50%	75.93%	2.26%	0.27%

What is probably needed in Hong Kong is a new public-engagement strategy in which the public is not merely passively consulted but is proactively engaged in the planning process. As underscored by Lidskog (1997), dialogue alone does not necessarily guarantee that an intended siting can be carried out. What matters most are frequent, open, and continuous interactions between the authority and the local community that describe options and alternatives, clarify interests, and aim at consensus building. Particular emphasis should be placed on risk communication and clearing up misunderstandings.

Our surveys (CEPRM, 2007a, 2007b) suggest that economic compensation is of secondary importance compared to other measures of enhancing public acceptance, such as adoption of effective public-consultation programs, environmental monitoring and safety audits, implementation of mitigation measures, and giving due consideration to options and project need. This is consonant with some previous findings (e.g., Jenkins-Smith & Kunreuther, 2005) that economic compensation is a promising measure only for low-risk and benign LULUs, such as landfills and prisons. In the case of Tuen Mun, improvements in transport facilities and community facilities are probably more effective measures than monetary rewards. Such improvements could enhance the image of Tuen Mun and counteract the labelling effect. It is probably because of the negative "label" that some local residents are resistant to having more LULUs in their district. Removing this stigma may be useful in enhancing the public acceptance of LULUs. Another possible approach is to undertake a comprehensive analysis of the need for LULUs of various types, to communicate with the stakeholders and allocate them strategically to various districts so that not a single district will feel that they are the only one to bear the burden of the society. This conclusion calls for a reexamination of strategies adopted and a move away from seeing siting merely as a technical issue to one that embraces the social, economic, and political dimensions.

CONCLUSIONS

The NIMBY syndrome poses significant challenges to development in Hong Kong, as it does in other countries. This phenomenon is particularly exacerbated by Hong Kong's small size and hilly terrain, and its small and administration-led government that accepts little public input from the local community in the planning process. It is paradoxical to observe that the residents of Tuen Mun, a district seen by many people as having a disproportionate share of LULUs, are not keenly aware LULUs in their district. This can be ascribed to the survey finding that not many people know about the planning/siting process involving LULUs, and the majority do not have a high level of trust in the government or the local district councils. In the absence of trust and a consultation process in which the public have faith, the majority of the public learn about LULUs through the news media and see LULUs as "add-ons" which may degrade the attractiveness of the community. To resolve the conflicts arising from the siting of LULUs, it is imperative to build trust and engage the key stakeholders and the public in meaningful dialogues focusing on risk communication, justification of the need for the project, and presentation of options and alternative alignments. The survey results indicate that monetary compensation is probably of limited effectiveness in reducing public resistance.

REFERENCES

Baxter, J. W., Eyles, J. D., & Elliott, S. J. (1999). From siting principles to siting practices: A case study of discord among trust, equity and community participation. *Journal of Environmental Planning and Management, 42*(4), 501–525.

Census and Statistics Department (CSD) (2007). *Thematic report: Household income distribution in Hong Kong*. Census and Statistics Department, The Hong Kong SAR Government. Retrieved April 15, 2009 from www.bycensus2006.gov.hk/FileManager/EN/Content_962/06bc_hhinc.pdf

Centre for Environmental Policy and Resource Management (CEPRM). (2007a, June 16). *Siting locally unwanted facilities in Hong Kong—public survey results*. CEPRM, Department of Geography and Resource Management, The Chinese University of Hong Kong. Retrieved April 15 2009 from ceprm.grm.cuhk.edu.hk/LULU/Workshops/LULUWorkshop_JL_Eng.pdf

CEPRM. (2007b). *Siting locally unwanted facilities in Hong Kong—Tuen Mun survey results*. CEPRM, Department of Geography and Resource Management, The Chinese University of Hong Kong. Retrieved April 15, 2009, from ceprm.grm.cuhk.edu.hk/LULU/Surveys/TM_survey_081016.pdf

Chiou, C. T. (2005). NIMBY syndrome and facility siting. *The Chinese Public Administration Review, 14*(3), 33–64.

Groothuis, P. A., & Miller, G. (1997). The role of social distrust in risk-benefit analysis: A study of the siting of a hazardous waste disposal facility. *Journal of Risk and Uncertainty, 15*(3), 241–257.

Jenkins-Smith, H. C., & Kunreuther, H. (2005). Mitigation and benefits measures as policy tools for siting potentially hazardous facilities: Determinants of effectiveness and appropriateness. In S. H. Lesbirel & D. Shaw (Eds.), *Managing conflict in facility siting: An international comparison* (63–84). Cheltenham, UK and Northampton, MA: Edward Elgar.

Johnson-Cartee, K. S., Graham B. A., & Foster, D. (1992/1993). Siting a hazardous waste incinerator: Newspaper risk communication and public opinion analysis. *Newspaper Research Journal, 13/14*(4/1), 60–72.

Kuhn, R. G., & Ballard, K. R. (1998). Canadian innovation in siting hazardous waste management facilities. *Environmental Management, 22*(4), 533–545.

Lai, P. W., Woo, L. Y., Lam, K. C., Lee, W. Y. , & Fung, T. (2007). *Siting and community response to locally unwanted land uses: A literature review*. Centre for Environmental Policy and Resource Management, Department of Geography and Resource Management, The Chinese University of Hong Kong.

Lam, K. C., & Brown, L. (1997). EIA in Hong Kong: Effective but limited. *Asian Journal of Environmental Management, 5*(1), 51–66.

Lesbirel, S. H., & Shaw, D. (2005). *Managing conflict in facility siting: An international comparison*. Cheltenham, UK and Northampton, MA: Edward Elgar.

Leverett, B., Hopkinson, L., Loh, C., & Trumbull, K. (2007a). Leadership. In B. Leverett, L. Hopkinson, C. Loh, & K. Trumbull (Eds.), *Idling engine: Hong Kong's environmental policy in a ten-year stall, 1997–2007* (77–84). Hong Kong: Civic-Exchange.

Leverett, B., Hopkinson, L., Loh. C., & Trumbull, K. (2007b). Case study on the Environmental Impact Assessment Ordinance (EIAO). In B. Leverett, L. Hopkinson, C. Loh, & K. Trumbull (Eds.), *Idling engine: Hong Kong's environmental policy in a ten-year stall, 1997–2007* (131–145). Hong Kong: Civic-Exchange.

Lidskog, R. (1997). From conflict to communication? Public participation and critical communication as a solution to siting conflicts in planning for hazardous waste. *Planning Practice & Research*, *12*(3), 239–249.

Macintyre, D., & Tashiro, H. (2000). Japan's dirty secret. *Time*, *156*(21). Retrieved April 15, 2009 from www.time.com/time/asia/magazine/2000/0529/japan.toxic.html

Ng, M. K. (2004). Sustainable development and planning. In T. Mottershead (Ed.), *Sustainable development in Hong Kong* (293–322). Hong Kong: Hong Kong University Press.

Popper, F. J. (1981). Siting LULUs. *Planning 47*, 12–15.

Popper, F. J. (1987). LP/HC and LULUs: The political uses of risk analysis in land-use planning. In R. W. Lake (Ed.), *Resolving locational conflict* (275–287). Piscataway, NJ: Centre for Urban Policy Research, Rutgers University.

Quah, E., & Tan, K. C. (2002). *Siting environmentally unwanted facilities: Risks, trade-offs and choices.* Cheltenham, UK and Northampton, MA: Edward Elgar.

Rahardyan, B., Matsuto, T., Kakuta, Y., & Tanaka, N. (2004). Residents' concerns and attitudes towards solid waste management facilities. *Waste Management*, *24*, 437–451.

Slovic, P. (1987). Perception of risk. *Science*, *236*(4799), 280–285.

Upreti, B. R., & van der Horst, D. (2004). National renewable energy policy and local opposition in the UK: The failed development of a biomass electricity plant. *Biomass and Bioenergy*, *26*(1), 61–69.

Wolsink, M. (1994). Entanglement of interests and motives: Assumptions behind the NIMBY-theory on facility siting. *Urban Studies*, *31*(6), 851–866.

Yoo, H. (1996). Siting of nuclear power plants in Korea. In D. Shaw (Ed.), *Comparative analysis of siting experience in Asia* (101–114). Taibei: Institute of Economics, Academia Sinica.

9

Community-Driven Regulation, Social Cohesion, and Landfill Opposition in Vietnam

Nguyen Quang Tuan and Virginia Maclaren

INTRODUCTION

The success of economic development in Vietnam since the adoption of *doi moi* in 1986 has not been without cost, notably the increasing environmental degradation of the country's air, land, and water resources (MONRE, 2006). In recognition of this problem, the Vietnamese government has passed a number of environmental regulations and has attempted to increase the capacity of environmental agencies. However, the regulatory regime continues to be weak. At times, legal regulations (such as Environmental Impact Assessment regulations) simply play a ritual role rather than find application in mitigating pollution successfully (Doberstein, 2003; O'Rourke, 2004). Lack of enforcement is a significant problem for a wide variety of environmental regulations (Dang & Nguyen, 2006). In light of these weaknesses, vocal opposition by communities has emerged as an important method for pressuring polluting facilities to reduce their pollution levels (Phuong & Mol, 2004; O'Rourke, 2004).

Recently, several researchers have examined the effectiveness of community pressure on polluting firms. Their studies reveal some promising strategies for responding to the environmental impacts of industrialisation and urbanisation in the presence of a lax environmental regulatory regime (O'Rourke, 2004; Phuong & Mol, 2004). Some refer to this type of pressure as "community-driven regulation" (CDR) (O'Rourke, 2004) while others use the term "informal regulation" (Hettige et al., 1996; Phuong & Mol, 2001). Although the evidence is limited to date, the outcome of community pressure appears to vary from case to case, depending on the social and economic characteristics of the affected communities, including social cohesion and external linkages (O'Rourke, 2004).

An important gap in the literature on CDR and informal regulation is that there has been no research on the issue of community activism against noxious public facilities. The main focus of previous research has been on community actions against industrial pollution. For a number of reasons, community actions against noxious public facilities might be different from those against industrial firms in Vietnam. First, unlike some public facilities such as landfills, industrial firms usually bring substantial economic development opportunities to local communities. In contrast, noxious public facilities like landfills are more likely to bring stigmatisation than improve employment prospects. The media also tends to play a different role in public versus private facility conflicts. As noted by O'Rourke (2004), media in Vietnam often side with local residents in their fight against polluting private firms. In contrast, residents who oppose publicly owned facilities, such as landfills or even state-owned enterprises, may receive less support from the media because the media is strictly controlled by the government.

This chapter examines the extent and effectiveness of community-driven regulation in dealing with landfill problems in Vietnam. It has several objectives, including characterisations of (1) the nature and effectiveness of CDR around landfills in Vietnam, (2) the factors that play a role in community activism against landfills, and (3) the influence of CDR and social cohesion on decision makers at higher levels of government and on regulators.

In discussing the effectiveness of communities in dealing with local industrial pollution, O'Rourke (2004) focused mainly on the ability of a community to reduce pollution and close a facility. In this study, we expand on the idea of an effective community in more detail, defining effectiveness relative to outcomes for the host communities rather than for society in general. Probably the most effective result of community opposition to a polluting facility that a community can achieve is to close the facility. The second-most effective is to allow the facility to continue its operation, but the facility must be well managed and produce zero or an "acceptable" level of pollution for local communities. If the facility is not closed or the pollution is not stopped or reduced, the ability of a community to force the government to provide more compensation is also a measure of community effectiveness. Compensation could take the form of direct monetary compensation or in-kind compensation, such as better local infrastructure or new employment opportunities. Finally, forcing government into dialogue is another measure of effectiveness, even though it may not necessarily result in any action by the government.

COMMUNITY-DRIVEN REGULATION

According to O'Rourke (2001, p. 124), community-driven regulation occurs when "communities directly pressure firms to reduce pollution, monitor industrial facilities, prioritise environmental issues for state action, pressure state environmental agencies to improve their monitoring and enforcement capacities, and raise public and elite awareness of environmental issues and trade-offs between development and environment." Other researchers have defined informal regulation in a similar manner (Pargal & Wheeler, 1996; Hettige et al., 1996; Phuong & Mol, 2001).

The main idea of CDR or informal regulation is that when formal regulation is weak or absent, communities often use other channels to achieve pollution abatement at local factories (Pargal & Wheeler, 1996). Without recourse to legal enforcement of existing regulations, communities must rely on their own efforts to challenge the polluter. Community pressure for abatement of pollution from factories can take many forms, including demands for compensation, complaints to authorities, threats to market reputation, social or political pressure on plant managers and regulators, proposals for negotiated pollution control agreements, civil disobedience, or even the acts of violence (O'Rourke, 2004; Hettige et al., 1996).

Case studies in many developing countries (e.g., Indonesia, China, Thailand, Bangladesh) have shown that community pressure is sometimes able to impose a cost that forces the polluting firm to change its level of pollution (Hettige et al., 1996). The greater the perceived damage and the community's ability to organise, the higher the compensation exacted by the community (Pargal & Wheeler, 1996). These case studies have also shown that communities may be able to pressure firms successfully to abate even in the absence of formal regulations or their enforcement.

Research in Vietnam has demonstrated similar findings to those from other developing countries: CDR in Vietnam has to some extent been successful in making polluting firms reduce their levels of pollution (Roodman, 1999; O'Rourke, 2004; Phuong & Mol, 2004). Using six case studies in different provinces of Vietnam, O'Rourke (2004) found that the success of CDR in pressuring industries and the state depended on social cohesion, connectedness, and capacity of the community. An important implication of CDR is that it may actually strengthen the authority of state environmental agencies because of public demands for environmental inspections and regulation enforcement. O'Rourke (2004) also notes that the economic dependency of communities living in a polluting firm's vicinity may reduce the effectiveness of CDR. However, several other case studies in Vietnam indicate that even when community members depend on industrial firms, they tend to react strongly and ask for better environmental quality when they are affected by pollution (e.g., Frijns et al., 2004; Phuong & Mol, 2004).

There has been little investigation of the relationship between the social and economic characteristics of local communities (e.g., income, occupation, and education levels) and the effectiveness of CDR. Examining the effectiveness of informal regulation in Indonesia, Hettige et al. (1996) found that poor communities with low levels of education and information might permit inappropriately high pollution, either because they are not aware of it, they cannot evaluate its consequences, or they are unable to organise to combat it. O'Rourke (2004) found similar results in Vietnam, in that poor communities were unable to organise as effectively as wealthier communities. Further research will help to confirm whether education and income are indeed related to CDR effectiveness in the context of Vietnam. Finally, as mentioned earlier, CDR has been applied to industrial firms in Vietnam and not to public facilities like landfills; therefore, the question of whether the elements of CDR are the same in this context also needs further investigation.

Adapting O'Rourke's concept of CDR for the present study, we redefine CDR as follows: CDR consists of various forms of community pressure on state and private facilities to reduce or stop the pollution from those facilities. We use the term "community-driven regulation" rather than "informal regulation" because it not only focuses on community initiatives, but it also represents the conjunction between communities and state agencies. CDR is not a static feature but may increase in intensity and change its form over time. For example, community activism against pollution may escalate and intensify when the demands of the community members remain unanswered.

The strength of community activism against a polluting facility has often been characterised by network density (i.e., the number of social networks in the community and the number of residents participating in them) (Kousis, 1999). For example, in his study of rural protests against a polluting facility in China, Jing (2003) found that the strength of family networks played an influential role in the village's struggle against pollution. The pervasiveness of pollution also affects the strength of community activism (Kousis, 1999). As distance from a polluting facility increases, there tends to be less exposure to pollution from that facility and less community activism. Elliott et al. (1993) found that actions taken against two landfills in Ontario, Canada were more likely to occur in closer proximity to the landfills. Although the regulatory and enforcement context is quite different from Vietnam, Elliott et al.'s (1993) findings that not only distance but also site, membership in an environmental group, and reciprocity were statistically significant in explaining the likelihood of taking action against a landfill, suggest that those variables may be worth exploring in the Vietnamese context as well.

The effectiveness of CDR measures in dealing with local industrial pollution in Vietnam appears to depend on a number of factors. Third parties such as the media are important to the success of CDR because they may play a role in uncovering and publicising causes of industrial pollution (Roodman, 1999; O'Rourke, 2004). However, as noted earlier, the question of whether a third party is willing to side with local residents opposing public facilities needs more investigation. O'Rourke (2004) indicates that communities face different challenges in mobilising against state-owned as opposed to private enterprises because the state, as owner, is often both polluter and regulator. Social cohesion may be another important factor affecting the effectiveness of CDR since community mobilisation often encounters collective action problems, shirking, and free-riders (O'Rourke, 2004; Rydin & Pennington, 2000).

In summary, previous research suggests that a number of variables may be influential in driving CDR. These include distance from the facility, social cohesion, socioeconomic characteristics of residents, site, and experienced impacts. The next section elaborates on the nature and contribution of social cohesion and social capital to CDR.

THE IMPORTANCE OF SOCIAL COHESION AND SOCIAL CAPITAL

Previous research suggests that social cohesion is a central element of successful CDR. This section elaborates on the concept of social cohesion and relates it to social capital. Social capital and social cohesion are closely linked in the literature. Some researchers consider social capital to be a part of social cohesion. For example, Kearns and Forrest (2000) argue, "A great constituent of social cohesion is common values, social order and control, social solidarity and reduction in wealth disparities, social networks and social capital." Other researchers consider social capital the same as social cohesion, which, in turn, is a function of factors, such as traditional interactions and institutions, a common heritage, values, etc. (Pargal et al., 1999). Berger-Schmitt (2002) defines social cohesion as having two dimensions: inequality and social capital. Inequality concerns the goal of promoting equal opportunities and reducing disparities and divisions within a society. This also includes social exclusion. The latter concerns the goal of strengthening social relations, interactions, and ties. From Berger-Schmitt's view, social cohesion seems to be a broader concept than social capital. Other researchers considering the distinction between social cohesion and social capital contend that social cohesion focuses on a sense of belonging (Jenson, 1998) and a sense of attachment to neighbourhoods (Stafford et al., 2003) more than social capital does. According to Stafford et al. (2003), social cohesion has two aspects: structural and cognitive. Structural aspects of social cohesion include family and friendship ties, and participation in organised associations. Cognitive aspects of social cohesion involve trust, attachment to neighbourhood, and practical help (or norms of reciprocity). Stafford et al.'s interpretation of the different types of social cohesion is similar to the interpretation of social capital given by Grootaert and van Bastelaer (2001).

Clearly, there is much overlap between social capital and social cohesion, and the difference between the concepts is still subject to debate. Almost all researchers agree that both social cohesion and social capital contain many of the same elements such as trust, networks, and norms of reciprocity (e.g., Lyon, 2000; Pretty & Ward, 2001). In this paper, we draw on Lavis and Stoddart (1999) in defining social cohesion as the networks, norms of reciprocity, and trust that bring people together to take action. The focus is on the quality and quantity of social interactions that appear in a community, rather than on the resources gained through these interactions (Stafford et al., 2003). This definition of social cohesion also implies cognitive aspects of social cohesion: a sense of attachment. Place attachment is particularly important in Vietnamese villages because of village history. Villages were often founded by a small group of people who were then worshipped by their descendents (Huy, 1993). The people in a village not only have close relationships because of their attachment to place, but also because of the historical relations of mutual assistance and solidarity from struggling with environmental uncertainties and foreign invasion (To, 1993).

Another reason why this definition seems appropriate is its inclusion of family and friendship ties (often referred to as strong ties). According to Dalton et al. (2002), family relationships are a central part of social capital in Vietnam, in which the role of parents is essential to the family. Dalton et al. further argue that the belief of the Vietnamese people in filial piety, the acceptance of patrilineal authority, and the traditions of ancestor worship have deepened the importance of family in Vietnamese society. Thus, research on social cohesion in Vietnam must take into consideration these two important characteristics. In addition, social cohesion in this paper can be understood as social capital combined with a sense of attachment to the neighbourhood.

The main perspective taken by many academics is that social capital/social cohesion is particularly important for individuals and communities with constrained material resources and is a precursor to collective action and potential community improvements (Ostrom, 1990; Lyon, 2000; Pretty &

Hugh, 2001; Fukuyama, 2001; Petro, 2001; Grootaert & van Bastelaer, 2001). Trust and networks are essential components of social capital and/or social cohesion which lubricate collective actions. According to Putnam (1993), the greater the level of trust within a community, the greater the likelihood of cooperation, and cooperation itself breeds trust. The denser the networks in a community, the more likely the community's members will be able to cooperate for mutual benefit (Coleman, 1988; Putnam, 1993). Thus, high levels of social capital are associated with cooperative social problem-solving because voluntary cooperation is easier in a community that has inherited a substantial stock of social capital (Coleman, 1988; Putnam, 1993).

While many researchers have written about the role of social capital in the sphere of economic and community development, few scholars, with the exception of Pargal et al. (1999) and Beall (1997), have examined the connection between social capital and waste management, and none have referred specifically to social cohesion.

One of the greatest weaknesses of social-capital and/or social-cohesion theory is the absence of consensus on how to measure it. A number of researchers have noted the challenges of measuring social capital (Putnam, 1993; Inkeles, 2000; Paldam, 2000). First, social capital consists of many social qualities that are difficult to quantify. Developing methods for measuring trust and networks is not easy and there is little consensus on how to do it (Petro, 2001). Second, the absence of a standard definition of social capital makes it difficult to develop indicators and methods for measuring social capital. Different types of associations can create different kinds of social capital (Coleman, 1988; Putnam, 1993; Petro, 2001). Finally, the complexity of social capital means that a single indicator cannot capture all of its aspects, although some researchers have used aggregation techniques to develop social capital indexes (c.f., Grootaert, 1999; Grootaert et al., 2002, Grootaert & Narayan, 2004).

Despite these problems, there are a number of variables that have been used frequently to measure the level of social cohesion and/or social capital in society, such as membership in voluntary associations (as a measure of social networks), the extent of interpersonal trust among citizens, and their perceptions of the availability of mutual aid (Putnam, 1993; Lochner et al., 1999). We use similar variables in this study.

METHODOLOGY

This paper investigates community-driven regulation in four communities in northern Vietnam which are adjacent to major landfills: Nam Son, Trang Cat,

Ha Khau, and Hung Dong communes. Residents in these four communities protested in a variety of ways and at various times against pollution from the landfills. A questionnaire survey was delivered to 730 randomly selected households in the four communes between January and May 2005. The number of questionnaires delivered in each commune was roughly proportional to the commune population (about 10 percent of households). The first part of the questionnaire included questions related to socioeconomic characteristics of the households and distance from the landfill. The second part asked about community concern about the landfill, such as perceived impacts. The third part explored information on CDR, including measures taken by the respondents to oppose the landfill, and the frequency and effectiveness of those measures. The final part addressed social cohesion of the community such as trust, networks, reciprocity, unity, and sense of place attachment. All of the questions were fixed-response questions, except the questions exploring reasons for the concern about the landfill and for the unity of the community, which were open-ended.

The overall response rate for the survey was high (76 percent) for a drop-off questionnaire, probably due, in part, to the high level of concern about the issue. Another reason for the high response rate was that the first author conducted two to three call-backs to nonresponding households in each study community. This helped raise the response rate considerably.

After the survey, follow-up interviews were chosen from a purposeful sample of about 10 residents at each site who had completed the survey and who were identified as being "active" participants in the landfill debate. An "active" participant was one who had concerns about the landfill and had made those concerns known in some way.

In addition to the follow-up interviews, 29 key informants were interviewed (in Vietnamese) at the study sites, including five environmental regulators from the provincial Departments of Natural Resources and Environment (DoNREs), four municipal waste management officials from the respective cities' Urban Environmental Companies (URENCOs), and twenty commune officials. DoNRE officials were selected because they are responsible for environmental monitoring and regulation of landfills, and URENCO officials are normally responsible for waste collection and for managing the landfill. Commune officials are often the most knowledgeable of all local residents about the siting and operations of the landfill.

The follow-up interviews and the interviews with officials and experts included questions of how the respondent and his/her organisation responded to community pressure to reduce the impacts from landfill operations; and whether his/her organisation considered factors such as strong social cohesion when deciding where to site the landfill. Most respondents understood the concept of social cohesion, but if they did not, it was explained during the interview.

RESULTS AND ANALYSIS

The results presented here first examine several aspects of the effectiveness of CDR, including our own assessment of effectiveness at the four sites, measures taken by respondents to oppose the landfill, and their satisfaction with the outcome of those measures. The second section looks at the relationship between community pressure and distance of respondents from the landfill while the third assesses the relationship with socioeconomic characteristics of the respondents. The fourth section presents results on the relationship between social cohesion and CDR. Several measures of social cohesion are examined, including trust, family, and friendship networks, associational activity, reciprocity, sense of unity in the community, and attachment to place. The final section looks at the relationship among CDR, social cohesion, and decision making by policy makers and regulators.

The Effectiveness of CDR

To analyse the effectiveness of CDR, it is useful to rank the study communities according to their success in dealing with landfill pollution. Table 1 identifies the outcomes of community pressure by site, based on interviews with commune, URENCO, and DoNRE officials. Overall, we judge Trang Cat to have been the most effective community in terms of both what it has achieved and what it has been promised. It was able to force the government to engage in dialogue with local residents. The government had to promise to close the landfill permanently within two years and manage the facility according to the national regulations for reducing environmental pollution. Since August 2004, the government has organised a number of meetings to persuade local residents to accept the government plan, but Trang Cat landfill remains closed since residents are still blocking the access road. Recently, the government has offered a new plan, which includes building a compost factory on the landfill site, mitigating the pollution, employing local workers in the compost factory, and creating a landfill-monitoring committee consisting of representatives from the community and commune officials. However, local residents have been reluctant to accept this new plan.

The Nam Son community was able to persuade the government to

Table 1. Results and Effectiveness of CDR by Site

Outcome of CDR	Trang Cat	Nam Son	Hung Dong	Ha Khau
Permanent closure of the landfill	Promised	Not requested	Promised	Not requested
Stopping/reducing pollution	Achieved by closure of landfill and promised when it opens again	Some achieved and more promised	Promised	Some achieved
Creation of a community monitoring committee	Promised	No	No	No
Provision of monetary compensation	Some achieved and more promised	Some achieved (about 40% of amount promised)	Some achieved (about 10–20% of amount promised)	No
Construction of a compost facility at the site employing local workers	Promised	Promised	Promised	No
Improved local infrastructure	Promised	Achieved	No	No
Forced government into dialogue	Achieved	Achieved	Achieved	No
Rank on effectiveness	1	2	3	4

provide better infrastructure, such as upgraded intervillage roads and a new primary school, provide more compensation, and according to a local official, was able to "satisfy about 40 percent of what the residents requested" for compensation. Hence, Nam Son is the second-most effective community. The Hung Dong commune was less effective than Nam Son because, according to a local official, the government "satisfied about 10 to 20 percent of what the residents requested" for compensation. Finally, the Ha Khau commune was the least effective because, according to a respondent from the follow-up interviews, there has been almost no change (since protests occurred in 2003 several months after the opening of the landfill) except a small reduction in pollution (smoke) from the landfill.

When asked whether they had used any measures to oppose the landfill, just over one-third (37 percent) of survey respondents said that they had. There was a statistically significant difference in the use of measures by site, with only 22.1 percent of Ha Khau residents reporting the use of measures, compared to 70.6 percent at Hung Dong (Table 2). Only about one-third of respondents at Trang Cat reported applying pressures of some form, which is surprising given the known record of extended activism at the site. In the discussion that follows, we use the terms "pressure" and "measures" interchangeably since pressure normally increases as the number of measures used increases.

The data in Table 2 classify respondents taking measures to oppose the facility by site and by their opposition to the facility. Not surprisingly, not everyone who claimed that they opposed the landfill did something about it. Only 51 percent of those who opposed the landfill took measures to pressure the authorities. This gap between concern and action is a common phenomenon in North America (c.f., Mansfield et al., 2001; Walsh & Warland, 1983). The percentage of nonactive but opposed respondents differed significantly across the study sites, with the highest percentage present at Trang Cat and Ha Khau.

Table 2. Community Pressure by Site and by Opposition

	Nam Son		Trang Cat		Ha Khau		Hung Dong		
	Pressure	No Pressure	Pressure	No Pressure	Pressure	No Pressure	Pressure	No Pressure	
Opposed	48 (53.9%)	41 (46.1%)	32 (33%)	65 (67.0%)	31 (41.3%)	44 (58.7%)	74 (72.5%)	28 (27.5%)	
Not Opposed	0 (0%)	10 (100%)	2 (14.3%)	12 (85.7%)	5 (5.7%)	83 (94.3%)	3 (42.9%)	4 (57.1%)	χ^2 = 81.361
Total	48 (48.5%)	51 (51.5%)	34 (30.6%)	77 (69.4%)	36 (22.1%)	127 (77.9%)	77 (70.6%)	32 (29.4%)	p = 0.000

The results of community pressure in Table 2 do not seem to have a positive relationship with the effectiveness rating in Table 1. We consider the Trang Cat community to be the most effective community, but it has a low percentage of residents applying pressure. As will be discussed later in the chapter, this might be due to the unwillingness of local residents to admit in the survey that they were involved in activism against the landfill. Contrary to Trang Cat, the Hung Dong community has a high percentage of residents applying pressure, but was ranked as next to last in effectiveness. The explanation may be that Hung Dong is an old dump with a long history of community opposition and pressure, and each time local pressure has led to dialogue with the government and promises by the government to close the facility. However, it has remained open. Therefore, the pressure has had little effect. Nam Son was ranked ahead of Hung Dong in terms of effectiveness but it had a much lower percentage of residents admitting that they applied pressure. A reason for this apparent discrepancy could be that disruption of waste disposal at Nam Son caused waste to pile up in the streets of Hanoi, the national capital. This produced almost an immediate response by local authorities who promised to clean up the dump and increase compensation to local residents.

The respondents at all of the study sites used similar measures to oppose the landfill, including writing letters to the government and media, as well as talking to community leaders, outside influential people, and landfill managers, and speaking up at village/ward meetings. At all sites, except Ha Khau, residents of the community blocked waste trucks. Almost four-fifths of the respondents who were opposed reported that they spoke to their commune leaders about their opposition to the landfill (Figure 1). Often, local residents started to oppose the facility by "talking," especially expressing problems to their commune leaders or in the village meetings. The opposition then escalated to "writing" such as writing letters to the government and media, and then to "acting" such as blocking waste trucks. "Talking," "Writing," and "Acting" are a three-step strategy of opposition; when local residents were not satisfied with their opposition outcome, they would move to the next level of opposition. However, fewer people move to the next level; and fewest moved to the highest level of opposition (or said that they did). One follow-up respondent complained that "We were talking a lot to the commune's leaders, talking a lot in the village meetings, and writing a lot of letters to governmental organisations, but nothing has been changed so that we had to block waste trucks."

A somewhat surprising result from Figure 1 is the low percentage of respondents who reported that they engaged in civil disobedience by blocking trucks. For example, only 3 of the 120 respondents from Trang Cat said that they blocked trucks, but according to local officials, there were hundreds of women and elderly people who participated in blocking the waste trucks at the landfill in August 2004.

One explanation for the low numbers is that 73 percent of the respondents who answered the questionnaire were male while those who participated in blocking trucks were mainly women and older residents. The use of women and older residents to block waste trucks was an important strategy for opposing the facility. A commune official of Trang Cat Commune described how "during the first days of community opposition to the landfill, there were men and women participating in every event. However, after managing to release two young men from a police van, local residents used only women and older men to block the landfill."

Figure 1. Measures Taken to Apply Pressure

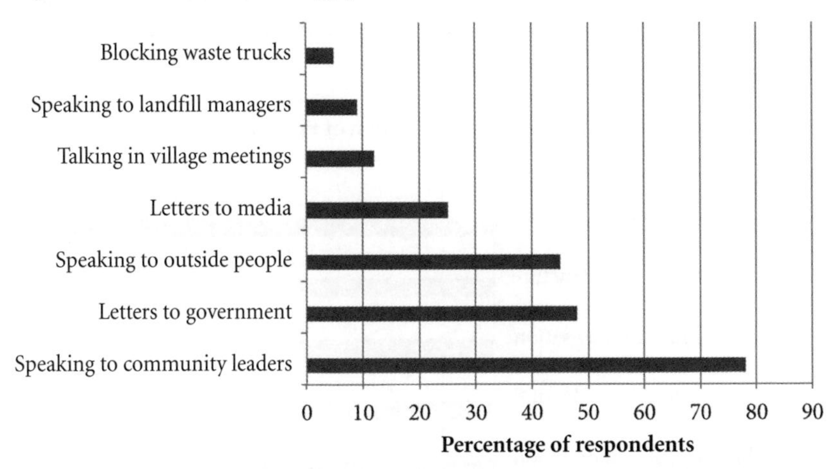

Another likely explanation is that Vietnam's Law on Environmental Protection (LEP) says that local residents have the right to complain about pollution, but it does not say they have the right to engage in civil disobedience. Therefore, blocking waste trucks might be considered an illegal act and respondents may have been reluctant to self-identify themselves or their family members as doing anything that was possibly illegal. A respondent in Trang Cat said that "if you conducted your survey before August 2004, before the local strike against the landfill, many more people would be willing to respond to the questions." In other words, opposition to the landfill has become a sensitive issue, at least in Trang Cat. This may also explain the relatively high percentage of respondents (67 percent) in Trang Cat who were willing to state that they were opposed to the facility but claimed that they had not taken measures to express that opposition.

Even local commune leaders at Trang Cat seem to have little understanding of residents' rights under the LEP since they asked the first author for a copy of the LEP and national regulations on landfill siting and operations for their commune people's committee. One of these officials said, "We want our residents to oppose the landfill legally, but we and our residents do not understand the main content of the LEP and national regulations on landfills."

To some extent, residents surrounding the four landfills believe that community application of pressure on the government has led to noticeable improvements in landfill management, as shown in Figure 2. However, more than 40 percent of the respondents who took measures to oppose the landfill felt that there were no significant changes in the landfill situation as a result of their actions. The most frequently cited change was better daily soil coverage; however, this change was noted by less than one-third of the respondents who took measures to oppose the facility.

Figure 2. Perceived Outcome of the Application of Pressure

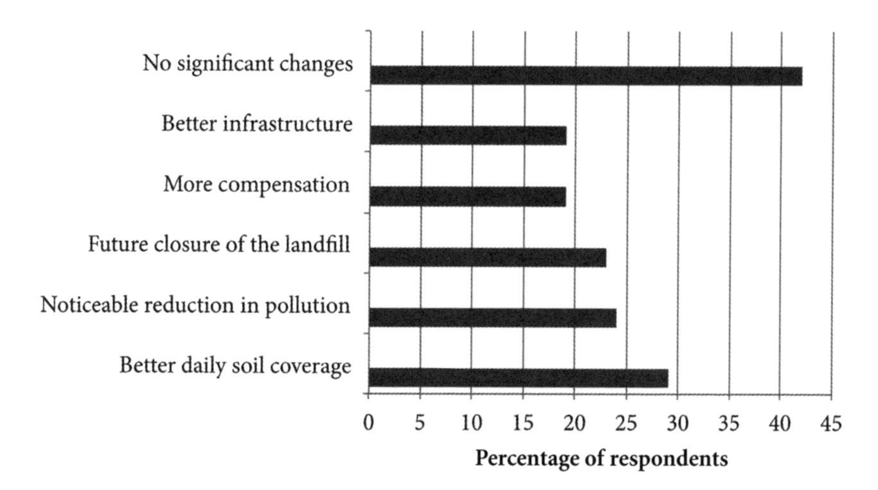

Table 3 presents the respondents' perceptions of the overall effectiveness of their activism by site and our rating of the perceived level of effectiveness (ranging from very low to medium) based on the percentage of respondents in each study community answering that a particular result had been achieved. The percentage of respondents saying that there had been specific improvements was no greater than about 60 percent at any site and was as low as 5 percent for the expectation that the landfill would be closed in the future and 8 percent for improvements in monetary compensation, both at Ha Khau. The percentage of respondents expressing satisfaction with the overall outcome of their efforts to oppose the landfill did not differ significantly by site and was generally low, ranging from 10 percent (at Nam Son) to 26 percent (at Ha Khau). The lack of difference in satisfaction among residents is surprising given the quite different outcomes that were achieved at the four sites.

Table 3. Perceived Results of Applying Pressure by Site*

	Nam Son	Trang Cat	Ha Khau	Hung Dong	χ^2, p-value
Better daily soil coverage	18 (37.5%) M	14 (38.9%) M	14 (36.8%) M	13 (16.5%) L	$\chi^2 = 10.480$ p = 0.015
Noticeable reduction in pollution	14 (29.2%) L	10 (27.8%) L	11 (28.9%) L	14 (17.7%) L	$\chi^2 = 3.152$ p = 0.369
Future closure of the landfill	5 (10.4%) L	10 (27.8%) L	2 (5.3%) VL	29 (36.7%) M	$\chi^2 = 19.957$ p = 0.000
More monetary compensation	15 (31.2%) M	4 (11.1%) L	3 (7.9%) VL	17 (21.5%) ML	$\chi^2 = 9.335$ p = 0.025
Better infrastructure provision	16 (33.3%) M	7 (19.4%) ML	5 (13.2%) L	10 (12.7%) L	$\chi^2 = 9.354$ p = 0.025
No improvements	19 (39.6%) M	17 (47.2%) M	18 (47.4%) M	30 (38.0%) M	$\chi^2 = 1.492$ p = 0.684
Satisfaction with outcome of activism	5 (10.4%) L	8 (22.2%) L	10 (26.3%) L	16 (20.3%) L	$\chi^2 = 3.859$ p = 0.277

* VL = very low level of effectiveness, L = low, ML = medium low, M = medium

There were statistically significant differences by site for respondents' perceptions of improved daily soil coverage at the landfill, the future closure of the landfill, monetary compensation, and better infrastructure provision. There were no differences by site in respondents' perception of a reduction in landfill pollution and overall perception of changes. A notable result is that residents of Nam Son were more likely than residents of other sites to feel that their actions had resulted in better infrastructure provision and improved monetary compensation. This result seems consistent with the comment by a local official that Nam Son residents have received about 40 percent of what the government had promised them, more than at the other landfill sites. Hung Dong residents were least likely to feel that there is now better daily soil coverage or a noticeable reduction in pollution, although they are happier with the compensation and infrastructure improvements. At the same time, Hung Dong had the highest percentage of respondents believing that the landfill will be closed in the near future.

Surprisingly, the Trang Cat community has successfully blocked the landfill for months, but has a lower percentage of respondents than Hung Dong, believing the promise from government that the landfill will be closed. Perhaps many residents in Trang Cat do not believe that they can "win" against the government since the "official" proposed closure date of the facility, when it is expected to reach its capacity, is 2020 while the official closure date is 2008 at Hung Dong. Trang Cat residents may also be discouraged by the perception that, even after many months of blocking access, the Hai Phong government is persistent in its negotiations to keep the landfill open. One resident commented in the follow-up interviews, "We think that we have not been successful yet. The success would come upon the government finding a new site for the city landfill and closing Trang Cat landfill."

Some of the responses by residents about the effectiveness of their actions against the landfill seem to contradict our "objective" assessment of the success of CDR in Table 1. There is no statistically significant difference among the sites in terms of perception of overall improvements, while we had expected respondents at Trang Cat to be most satisfied with the effect of their actions. The difference may be due to a lack of faith by Trang Cat residents in promised changes and their frustration with having to continue blocking trucks. In other words, satisfaction with the outcome of CDR efforts may only occur when there are specific gains or achievements rather than promises. One respondent noted, "We do not know how long we will be able to block the landfill. The government is very patient. One day we may have to accept their plan as the number of residents accepting the government plan is increasing." Another explanation is that measures taken to oppose the landfill brought both positive and negative impacts to the Trang Cat community. Several residents interviewed during the follow-up survey blamed the presence of many policemen in their community on opposition to the facility. Other respondents said that the presence of policemen negatively affected their daily businesses. One follow-up respondent commented, "I can't stand seeing policemen wandering all day and night in our commune." Another respondent felt even more strongly, saying, "I hate them [policemen] as much as I hated American soldiers before. They look at us as if we were the rebels."

Table 4 presents the results of the respondents' evaluation of the effectiveness of individual measures taken to oppose the landfill. The most effective measure, namely writing letters to government, was considered effective by less than half of the respondents. Although respondents rated the effectiveness of letters to the media as second highest after letters to government, about two-thirds of respondents felt that appealing to the media was ineffective. Although the media appears to play an important role for the success of communities in fighting industrial pollution (O'Rourke, 2004; Phuong & Mol, 2004), the media seems unwilling to side with local residents in fighting landfill pollution. A possible explanation for this reluctance is that landfills are owned by local governments, the media are state-controlled, and despite the individual's right to complain about pollution enshrined in the LEP, criticism of landfill operations by the media might be interpreted as criticism of the state. Many participants in the follow-up interviews claimed that they sent numerous letters to the media about pollution, but received no response. Often the media reports on landfill issues only when the consequences of opposition are impossible to ignore, such as reporting on community opposition at Nam Son landfill only after local residents blocked trucks and caused waste to pile up on the streets of Hanoi for several days. Commenting on media reports of misuse of funds at Trang Cat landfill, one commune official said that "they [media] just wanted to report on the issue of local corruption, but not landfill pollution." Corruption issues were probably of much more interest to Vietnamese readers—and the state—than the pollution issues at the landfill since corruption was the focus of statewide crackdown at all levels of government during the period of study.

Table 4. Perceived Effectiveness of Individual Measures Applied to Oppose the Facility

Measures	Effectiveness (number and % of the respondents answering the question)		
	Effective	Not effective	Total
Letters to the government	34 (42.5%)	46 (57.5%)	80 (100%)
Letters to media	18 (35.3%)	33 (64.7%)	51 (100%)
Speaking to community leaders	31 (29.2%)	75 (70.8%)	106 (100%)
Speaking to outside people	12 (24%)	38 (76%)	50 (100%)
Talking in the village meetings	8 (27.6%)	21 (72.4%)	29 (100%)
Blocking waste trucks	1 (11.1%)	8 (88.9%)	9 (100%)

O'Rourke (2004) claims that connections to influential outsiders have been seen as helpful in fighting industrial pollution in Vietnam. However, the residents of the four study sites did not feel that speaking to outsiders about their problems was effective. Possibly, since these are generally poor communities, the residents had few influential outsiders to call on. Their poverty and lack of access to influential outsiders could also have been a factor in each community's choice as a site for a landfill in the first place.

Talking in village meetings is one of the least-effective measures according to local residents. In a country with a strong central government like Vietnam, villages often have little influence on the decisions of the provincial and central government. This frustration about lack of power at the village level was expressed by one respondent who said, "We talk a lot in the village meetings, but nothing has been changed."

Distance Decay of CDR Measures

Did the use of measures decline by distance from the landfill site? Since taking action against the landfill requires an investment of time and effort, it might be expected that those who are closest to the landfill and experiencing the most severe impacts would be more likely to apply pressure on the government and the facility managers. Looking at all four sites together, there is a statistically significant decline in the application of measures by distance from the landfills (see Table 5).

Table 5. Distance Decay of Pressure by Site

	0–500m	501–1,000m	1,001–1,500m	>1,500m	χ^2, p-value
Nam Son					
Pressure	10 (83.3%)	18 (62.1%)	17 (53.1%)	3 (9.4%)	$\chi^2 = 27.705$
No Pressure	2 (16.7%)	11 (37.9%)	15 (46.9%)	29 (90.6%)	$p = 0.000$
Ha Khau					
Pressure	19 (59.4%)	9 (22.0%)	3 (10.7%)	6 (6.6%)	$\chi^2 = 43.991$
No Pressure	13 (40.6%)	32 (78.0%)	25 (89.3%)	85 (93.4%)	$p = 0.000$
Trang Cat					
Pressure	14 (38.9%)*		17 (28.8%)	4 (16.7%)	$\chi^2 = 3.445$
No Pressure	22 (61.1%)		42 (71.2%)	24 (83.3%)	$p = 0.179$

* Distance categories were combined ito meet the assumptions of the χ^2 test

	0–500m	501–1,000m	1,001–1,500m	>1,500m	χ^2, p-value
Hung Dong					
Pressure	32 (76.2%)	24 (66.7%)	8 (57.1%)	13 (68.4%)	$\chi^2 = 2.036$
No Pressure	10 (23.8%)	12 (33.3%)	6 (42.9%)	6 (31.6%)	$p = 0.565$
Four sites combined					
Pressure	66 (66.7%)	64 (47.4%)	45 (33.8%)	26 (15.7%)	$\chi^2 = 77.038$
No Pressure	31 (33.3%)	71 (52.6%)	88 (66.2%)	140 (84.3%)	$p = 0.000$

However, the distance decay effect is not present for all sites. Ha Khau and Nam Son exhibit distance decay, but Trang Cat and Hung Dong do not. These differences are hard to explain.

The hypothesis that those who are experiencing the greatest impacts from pollution are more likely to pressure the government turns out not to be relevant, except for Ha Khau, since at the other three sites, almost all residents report experiencing impacts, regardless of their distance from the landfill (see Table 6). Only at Ha Khau do application of pressure and experienced impacts decline with distance from the landfill. Another possible explanation for the differences in measures applied is that social cohesion might be particularly high in Trang Cat and Hung Dong, and high social cohesion could mean that residents are more likely to support their neighbours in taking action, regardless of where they live in the community. However, as will be shown later, social cohesion is low in Trang Cat and relatively high in Hung Dong.

Table 6. Distance Decay of Experienced Impacts by Site

Site	0–500m	501–1,000m	1,001–1,500m	>1,500m	χ^2, p-value
Nam Son	12 (100%)	29 (100%)	32 (100%)	30 (88.2%)	No decay
Trang Cat	7 (100%)	29 (100%)	58 (98.3%)	22 (91.7%)	No decay
Ha Khau	30 (93.8%)	26 (68.4%)	15 (60.0%)	17 (20.5%)	Decay present $\chi^2 = 59.571$ $p = 0.000$
Hung Dong	45 (97.8%)	36 (92.3%)	14 (100%)	18 (94.7%)	No decay

CDR and Socioeconomic Characteristics

Application of chi-square tests to check for relationships between the number of respondents who had applied measures to oppose the landfill and their social and economic characteristics reveals a relationship between occupation and community pressure, but not between pressure and either income or education (Table 7).

Table 7. Community Pressure and Socioeconomic Characteristics of Respondents

		Pressure	No pressure	χ^2, p-value
Occupation	Farmers	151 (77.8%)	169 (53.5%)	$\chi^2 = 30.500$
	Nonfarmers	43 (22.2%)	147 (46.5%)	p = 0.000
Education	Primary	11 (6.7%)	27 (10.5%)	$\chi^2 = 5.265$
	Basic	76 (46.6%)	95 (37.1%)	p = 0.153
	Secondary	62 (30.8%)	102 (39.8%)	
	University	14 (8.6%)	32 (12.5%)	
Income	Low	85 (64.4%)	151 (64.3%)	$\chi^2 = 1.722$
	Medium	34 (25.8%)	69 (29.4%)	p = 0.423
	High	13 (9.8%)	15 (6.4%)	

Farmers applied more pressure than nonfarmers, likely because landfill pollution into the surrounding farmland can have a negative impact on their livelihoods. One follow-up respondent at Nam Son landfill claimed that "we [the villager farmers] believe that the landfill pollution has an effect on reducing our agricultural productivity." This may explain why, in Ha Khau commune where the majority of residents are nonfarmers, the amount of pressure applied was lowest among the four sites.

CDR and Social Cohesion

We measured social cohesion on a variety of dimensions, including the strength of networks, trust, reciprocity, sense of unity, and attachment to the neighbourhood. In general, the measures of social cohesion used for this study point to the presence of high to very high levels of social cohesion at the four sites. In response to the question of whether a resident participates in any community or national associations, most respondents (86.8 percent) said that they participated in at least one association. Respondents are members of 2.79 associations, on average. Respondents at Ha Khau (mean = 2.53), Nam

Son (mean = 2.63), and Trang Cat (mean = 2.66) participate in the smallest number of associations. Respondents at Hung Dong (mean = 3.47) participate in a significantly, although perhaps not substantively, higher number of associations than do respondents at other three sites (based on a post-hoc comparison to all three sites using the Games-Howell statistic and $\alpha = 0.005$). The three most frequent groups that the respondents and their families participated in are the Vietnam Women's Union, the Farmers' Association, and the Youth Union (Figure 3). These associations are controlled by the state, making it difficult for local leaders of the associations to side with residents in opposing a public facility, but respondents from the follow-up interviews noted that the membership in associations was still beneficial in fighting the landfill because of the networking opportunities that they offered.

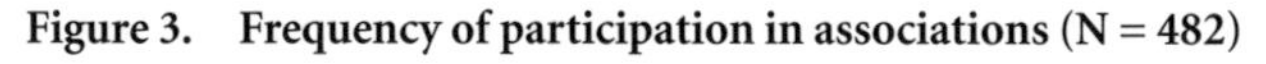

Figure 3. Frequency of participation in associations (N = 482)

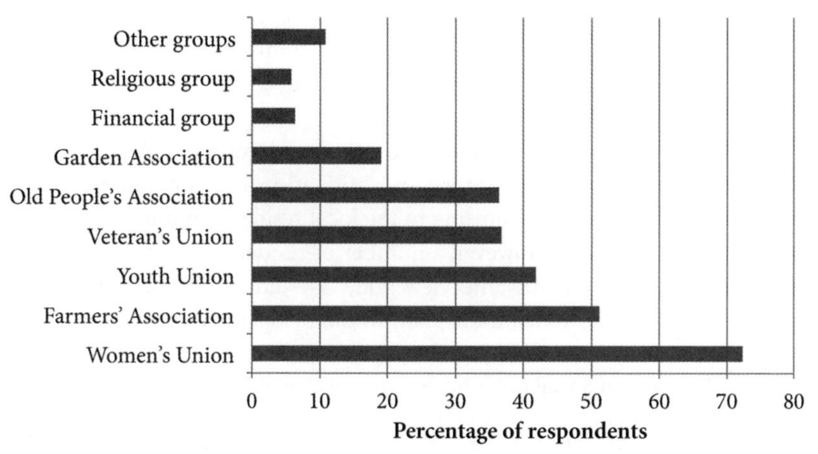

Sense of attachment to place was stronger than any other measure of social cohesion, with 96 percent of respondents indicating that they felt an attachment to their community (see Table 8). Some exceptions to the high levels of social cohesion were found in the measures of reciprocity, which ranged from extremely low (only 7 percent of respondents would seek help from their neighbours in looking after their children) to moderately high (63 percent of respondents would be able to count on their relatives for financial help and 60 percent could rely on relatives for child care). The higher level of reciprocity with relatives lends evidence to the importance of family networks in Vietnamese villages. Perhaps surprisingly, a larger percentage of respondents said that they could rely more on association members than on their neighbours for both financial help and child care. This could be due to the presence of existing programs for financial assistance and child care within one or more of the associations. Commune officials at the study sites admitted that members of associations such as the Women's Union, Farmers' Association, and Youth Union find it easier to obtain a bank loan than a person who did not participate in any association. For several measures of social cohesion, there were no statistically significant differences by site. These included the presence of close friends in the community, trust in association members, reliance on close friends for financial or child care help, and reliance on association members for financial help.

Surprisingly, the lowest level of trust and perceived unity in the community was found in Trang Cat—the community we had judged to be most effective in CDR. A possible explanation is that at the time of the survey, the situation in Trang Cat was tense: some women were still camping at the landfill and blocking waste trucks; and a number of policemen were present in the commune. Both officials and residents at Trang Cat agreed that opposition to the landfill caused some division within their community. Another divisive activity may be the ongoing negotiations with the city over a plan to reopen the landfill. According to a commune official at Trang Cat, the new proposal by the Hai Phong People's Committee to construct a compost plant at the site and introduce other compensatory and mitigation measures in return for reopening the site "seems promising to the local residents and many residents want to accept the plan." However, a number of residents doubted the plan. A respondent in the follow-up interview said, "We did not think that the plan could reduce the pollution and the plant would be operated as usual."

Table 8. Measures of Social Cohesion by Site and Overall

	Nam Son	Trang Cat	Ha Khau	Hung Dong	χ^2; p-value	Four sites combined
Presence of relatives and close friends in the village						
Relatives	88 (91.7%)	50 (78.1%)	78 (43.3%)	85 (94.4%)	$\chi^2 = 110.033$ p = 0.000	78%
Close friends	85 (78.7%)	82 (68.3%)	131 (64.2%)	84 (68.9%)	$\chi^2 = 6.957$ p = 0.073	69%
Trust						
Trust in village residents	50 (82.0%)	21 (38.9%)	63 (79.7%)	47 (87.0%)	$\chi^2 = 41.574$ p = 0.000	73%
Trust in relatives	65 (98.5%)	39 (72.2%)	88 (93.6%)	71 (94.7%)	$\chi^2 = 29.791$ p = 0.000	91%

	Nam Son	Trang Cat	Ha Khau	Hung Dong	χ^2; p-value	Four sites combined
Trust in close friends	62 (98.4%)	36 (66.7%)	97 (99%)	68 (90.7%)	$\chi^2 = 49.335$ $p = 0.000$	91%
Trust in neighbours	58 (90.6%)	35 (67.3%)	96 (90.6%)	56 (80.0%)	$\chi^2 = 17.020$ $p = 0.001$	84%
Trust in network members	33 (82.5%)	42 (91.3%)	72 (92.3%)	50 (92.6%)	$\chi^2 = 3.535$ $p = 0.316$	90%
Reciprocity						
Financial help from relatives	72 (66.7%)	89 (74.2%)	107 (52.5%)	78 (64.5%)	$\chi^2 = 16.768$ $p = 0.001$	63%
Child care help from relatives	81 (75.0%)	81 (67.5%)	100 (49.0%)	73 (59.3%)	$\chi^2 = 23.249$ $p = 0.000$	60%
Financial help from close friends	53 (49.1%)	56 (46.7%)	92 (45.1%)	68 (55.3%)	$\chi^2 = 3.388$ $p = 0.336$	49%
Child care help from close friends	55 (50.9%)	64 (53.5%)	112 (54.9%)	55 (44.7%)	$\chi^2 = 7.628$ $p = 0.108$	34%
Financial help from social network members	27 (25.0%)	37 (30.8%)	51 (25.0%)	24 (19.5%)	$\chi^2 = 4.148$ $p = 0.246$	25%
Child care help from social network members	11 (10.2%)	22 (18.3%)	20 (9.8%)	33 (26.8%)	$\chi^2 = 20.177$ $p = 0.000$	16%
Financial help from neighbours	7 (6.5%)	24 (20.0%)	14 (6.9%)	21 (17.1%)	$\chi^2 = 18.622$ $p = 0.000$	12%
Child care help from neighbours	3 (2.8%)	17 (14.2%)	3 (1.5%)	13 (10.6%)	$\chi^2 = 25.959$ $p = 0.000$	7%
Unity and sense of attachment						
Unity	81 (88.0%)	35 (47.9%)	86 (57.3%)	74 (87.1%)	$\chi^2 = 53.232$ $p = 0.000$	70%
Attachment	87 (100%)	52 (86.7%)	87 (96.7%)	76 (98.7%)	50% of expected cells less than 5	96%

Like Trang Cat, Ha Khau appears to have relatively low levels of social cohesion. Respondents from Ha Khau have the fewest relatives in their community. Reciprocity on four measures is lowest, or close to lowest in Ha Khau. Perceived community unity is also relatively low. The low sense of unity may be due to the lack of dense family networks in the community. This conclusion is consistent with the claim made by Dalton et al. (2002) that family ties are a central part of social capital in Vietnam. The low unity may also be due to the distinct division of occupations in the community.

Unlike Trang Cat, we had judged Ha Khau to have been the least successful with CDR. The survey results also suggested that Ha Khau had been least effective in a number of ways. Ha Khau had the fewest residents who applied measures to oppose the landfill and Ha Khau residents who took action were least satisfied with the outcome of several of those actions. Therefore, in contrast to Trang Cat, CDR at Ha Khau has not been as effective even though the two communities have similarly low levels of social cohesion.

As noted at the beginning of this section, almost all respondents at the four study sites expressed a strong sense of attachment to their communities. It was not possible to use a chi-squared test for identifying statistically significant differences in attachment by site because of the small percentage of respondents who felt no attachment.

CDR, Social Cohesion, and Decision Making

During the interviews with experts and officials, community pressure was found to be influential in environmental decision making. Upon being asked about the role of CDR, all officials and experts at the national and provincial levels admitted that expressed opposition was important for bringing the problems of local communities to the attention of higher levels of government. A DoNRE official said that "our staff is lacking personnel. Thus, we cannot cover everything related to the environment in our province. If there were no community opposition to the landfill, we would never know about the problem." Another DoNRE official said that "the [provincial] people's committee has to focus more on economic development and key economic indicators rather than on environmental protection. Strong community opposition to pollution is a good way to let the government know what is happening in the locality." Similarly, a key informant commented, "Community opposition to polluting facilities is now occurring everywhere in our country. Opposition to waste-disposal facilities like Trang Cat landfill in Hai Phong or Nam Son landfill in Hanoi has becomes a widespread concern in the country. Policy makers cannot keep silent; they have to respond quickly."

CDR seems to be able to influence decision-making processes by facilitating and improving the implementation of national regulations. As noted by Kunreuther (in this volume), the establishment of stringent standards for waste facilities and the enforcement of those standards once a facility has been opened are keys to addressing public perceptions about the risks posed by the facility. Similar to other developing countries, environmental agencies in Vietnam are facing many challenges from other governmental organisations in the design and implementation of environmental regulations (O'Rourke, 2004; Tang et al., 2005). Community actions help strengthen the position of the environmental department in municipalities. A DoNRE official admitted, "We can conduct regular inspections on pollution only once or twice a year. However, if we receive complaint letters from communities, we can conduct additional inspections." He further added that "communities to some extent help improve the quality of implementing laws because they want to see an actual reduction in pollution, not something written on documents like this or that is under the national environmental standards." In a country like Vietnam where most municipal leaders are thinking of economic growth and of meeting economic targets set by the national or provincial government, environmental inspections of polluting facilities might be interpreted as measures to prevent "development." Another DoNRE official commented, "We often face criticism from other departments for our 'active' inspections. These departments imply that the DoNRE, at times, is slowing down industrial production." Thus, CDR helps keep the balance of power between departments in the government.

Social cohesion was also found to have an influence on environmental decision making. DoNRE and URENCO officials in all four study sites felt that social cohesion provided communities with more power to bargain with polluting facilities and a bigger voice with the provincial and national governments. Social cohesion has provided local residents with more confidence and solidarity in fighting against pollution, as reflected by a comment from a commune official at Trang Cat who said that "the police can arrest one person, but cannot arrest the whole village."

Environmental policies in Vietnam are lacking, overlapping, and not meeting the demand for social and economic development of the country (CPV, 1998). A MONRE official admitted the weakness of existing environmental policies by saying that "our [environmental] policies respond very slowly compared to the fast-growing economy of the country."

One of the major weaknesses of environmental policies in Vietnam is the lack of public participation in both policy design and implementation. Public participation is difficult to realise in many developing countries where officials are not interested in encouraging representational democracy (Tosun, 2000). Public participation also requires considerable time, money, and an appropriate approach to be effective (Timothy, 1999; Tosun, 2000), so that it is not a high priority for bureaucrats. Social cohesion may facilitate participation processes (Rydin & Pennington, 2000) and allow local residents to have a bigger voice in calling for their rights to participate in environmental decision making.

CONCLUSION

CDR consists of steps taken by local communities to deal with their environmental problems and is characterised by an escalating strategy of "talking," "writing," and "acting" in opposition to the polluting facility. It involves a dynamic process developed from residents' attitude toward the polluting facility to their actions against the facility. However, CDR in the context of landfills appears to be somewhat different from CDR in the context of industrial pollution. While community opposition to industrial pollution has received much attention from the media (Roodman, 1999; O'Rourke, 2004; Phuong & Mol, 2004; Thanh, 2006), community activism against landfills has relied less on media support. Reasons for this include the sensitivity of landfill issues and the strict control of media by the government. Nor was speaking to influential outsiders felt to be helpful. Thus, the effectiveness of CDR in dealing with landfill problems mainly depends on communities' internal capacity.

Few residents felt that the measures they had taken to oppose the landfill were effective. In particular, there was little support expressed for the effectiveness of blocking truck access to the landfill. Despite the lack of support, blocking access appears to have forced the government into negotiating with the three communities that used this measure. Two of the communities, Hung Dong and Nam Son, were sufficiently satisfied with the negotiations that they stopped blocking access, even though they only received 20 to 40 percent of what they had asked for, respectively. O'Rourke (2004) claims that one of the constraints on the effectiveness of CDR in Vietnam is that even limited victories can appease residents' anger and have a demobilising effect among those opposing a facility. On the other hand, residents of Trang Cat have continued to block access even though they have been able to extract more concessions from the government than were offered at the other landfill sites. It is possible that the pervasiveness of the pollution, as suggested by Kousis (1999) in his study of community mobilisation in Europe, and the fact that the landfill is not expected to reach capacity until 2020 are what is driving the continued blockade. Although the livelihoods of the farmers in Trang Cat are particularly sensitive to leachate leakages from a landfill because of their reliance on aquaculture production, about an equal percentage of farmers and nonfarmers admitted to have taken measures to oppose the landfill.

Family networks were important for local residents in dealing with landfill pollution. Respondents also felt that membership in state-sponsored associations was beneficial in fighting the landfill because of the networking opportunities that they offered. With the exception of Trang Cat, the link between social cohesion and effectiveness of CDR seems to be positive. Ha Khau had low cohesion and low effectiveness. Nam Son and Hung Dong had relatively high cohesion and were more effective in their use of CDR than Ha Khau.

One might expect that rich communities opposing the polluting facility would be more active and effective than poor communities or communities with low education levels (Hettige et al., 1996). This study found no relationship between wealth or education and the likelihood that an individual would take measures to oppose the landfill. On the other hand, there was a statistically significant relationship between occupation of the respondents and community pressure. Farmers appeared to oppose the polluting facility more vigorously than the nonfarmer respondents, probably because landfill pollution puts their livelihoods at risk.

CDR appears to play a role in implementation of environmental regulations where enforcement of regulations is lax. Poor enforcement of regulations contributes to heightened public concern about the environmental risks of a landfill. CDR influences environmental decision-making processes by making decision makers at the provincial and national levels aware of local problems. Since the Vietnamese authorities are currently focused more on economic development than environmental protection, CDR can help balance the power between the environmental department and other departments in the government. However, fighting the landfill may also divide the community (e.g., Trang Cat community), especially when the government is very patient.

Although CDR can be beneficial in balancing power, it has its drawbacks. According to O'Rourke (2004), CDR does not work well in communities that are divided or poorly organised. In this study, we found that although Trang Cat was divided, it was effective as a community in acting against landfill pollution. This finding seems to contradict O'Rourke, but in fact, the divisiveness within the community appears to have been a result of residents taking action. In other words, CDR can itself produce divisions within communities. O'Rourke further notes that residents use CDR only against pollution problems that are clearly evident, such as those that they can taste or smell or see.

Hence, he stresses the importance of strengthening environmental agencies and their monitoring capacities.

One policy implication of this research is that more attention needs to be paid to landfill compensation practices. Research from North America has shown that compensation may be able to reduce opposition to waste facilities (see Kunreuther's chapter in this volume). Although monetary compensation was offered to residents in three of the case-study landfills, residents were not satisfied with the level of compensation, so its impact on changing attitudes towards the landfill was limited. Nguyen and Maclaren (2007) recommend that compensation for land expropriation at landfill sites in Vietnam should be made in negotiation with local residents and in accordance with market mechanisms rather than with prices imposed by the state. They also favour the use of nonmonetary compensation, such as provision of employment opportunities at the landfill or at complementary facilities at the site, such as the composting facility proposed by officials for Trang Cat in Hai Phong.

Another policy implication is that more formal mechanisms are needed to involve the public in siting and operations of noxious facilities. The proposal by the city government in Hai Phong to introduce a community monitoring committee at the landfill is one step in this direction. Public involvement in siting may result in the delay of a project, but could help to avoid the cost of community opposition in the future.

REFERENCES

Beall, J. (1997). Policy arena: Social capital in waste—a solid investment? *Journal of International Development*, *9*(7), 951–961.

Berger-Schmitt, R. (2002). Considering social cohesion in quality of life assessments: Concept and measurement. *Social Indicators Research*, 58, 403–428.

Coleman, J. S. (1988). Social capital in the creation of human capital. *American Journal of Sociology*, 94, S95–S120.

Community Party of Vietnam (CPV) (Dang Cong San Viet Nam). (1998). Chi thi 36/CT-TW cua Bo Chinh Tri ve tang cuong cong tac bao ve moi truong trong thoi ky cong nghiep hoa, hien dai hoa dat nuoc (Directives of the Politburo on enhancing environmental protection during the period of industrialization and modernisation of the country).

Dalton, R. J., Pham, M. H., Pham, T. N., & Nhu N. T. O. (2002). Social relations and social capital in Vietnam: Findings from the 2001 world values survey. *Comparative Sociology*, *1*(3/4), 369–386.

Doberstein, B. (2003). Environmental capacity-building in a transitional economy: The emergence of EIA capacity in Vietnam. *Impact Assessment and Project Appraisal*, *21*(1), 25–42.

Elliott, S. J., Taylor, S. M., Walter, S., Stieb, D., Frank, J., & Eyles, J. (1993). Modelling psychosocial effects of exposure to solid waste facilities. *Social Science and Medicine*, *37*(6), 791–804.

Frijns, J., Phuong, P. T., & Mol., A. P. J. (2004). Ecological modernization theory and industrializing economies: The case of Vietnam. In A. P. J. Mol & D. A. Sonnefeld (Eds.), *Ecological modernization around the world—perspectives and critical debates*. London: Frank Cass.

Fukuyama, F. (2001). Social capital, civil society and development. *Third World Quarterly*, *22*(1), 7–20.

Grootaert, C. (1999). Social capital, household welfare and poverty in Indonesia. Local level Institutions Study Working Paper No. 6, The World Bank, Washington, DC.

Grootaert, C., & Narayan, D. (2004). Local institutions, poverty and household welfare in Bolivia. *World Development*, *32*(7), 1179–1198.

Grootaert, C., Oh, G. T., & Swamy, A. (2002). Social capital, household welfare and poverty in Burkina Faso. *Journal of African Economics*, *11*(1), 4–38.

Grootaert, C., & van Bastelaer, T. (2001). *Understanding and measuring social capital: A synthesis of findings and recommendations from social capital initiatives*. The World Bank, Social Capital Initiative Working Paper No. 24, The World Bank. Retrieved from www.iris.umd.edu/adass/proj/soccappubs.asp

Hettige, H., Mailnul, M., Pargal, S., & Wheeler, D. (1996). Determinants of pollution abatement in developing countries: Evidence from South and Southeast Asia. *World Development*, *24*(12), 1891–1904.

Huy, V. (1993). A survey of the foundation of villages in Ha Nam, the traditional village in Vietnam. Hanoi: The Gioi Publishers.

Inkeles, A. (2000). Measuring social capital and its consequences. *Policy Sciences*, 33(3–4), 245–268.

Jenson, J. (1998). *Mapping social cohesion: The state of Canadian research*. Canadian Policy Research Networks Inc., CPRN Study No. F 03, Ottawa, Canada.

Jing, J. (2003). Environmental protests in rural China. In E. Perry & M. Selden (Eds.), *Chinese society: Change, conflict and resistance*. London and New York: Routledge.

Kearns, A., & Forrest, R. (2000). Social cohesion and multilevel urban governance. *Urban Studies*, *37*(5/6), 995–1017.

Kieu, M. (2005). Vi sao dan di to cao ve dat dai [Why do residents go for petitions on land-use]? Retrieved from www.vietnamnet.com.vn/xahoi

Kousis, M. (1999). Sustaining local environmental mobilizations: Groups, actions and claims in Southern Europe. *Environmental Politics*, *8*(1), 172–198.

Lavis, J. N., & Stoddart, G. L. (1999). *Social cohesion and health* (WP 99-09). Hamilton, Ontario: McMaster University.

Lober, D. J. (1995). Beyond self-interest: A model of public attitudes towards waste facility siting. *Journal of Environmental Planning and Management*, *36*(3), 345–363.

Lochner, K., Ichiro, K., Kennedy, B. P. (1999). Social capital: A guide to its measurement. *Health & Place*, *5*(4), 259–270.

Lyon, F. (2000). Trust, networks, and norms: The creation of social capital in agricultural economies in Ghana, *World Development*, *28*(4), 663–681.

Mansfield, C., van Houtven, G & Huber, J. (2001). The efficiency of political mechanisms for siting nuisance facilities: Are opponents more likely to participate than supporters? *Journal of Real Estate Finance and Economics*, *22*(2&3), 141–161.

Ministry of Natural Resources and Environment of Vietnam (MONRE) (Bo Tai Nguyen va Moi Truong, Vietnam). (2006). Bao cao Hien trang Moi truong Vietnam 2005. Retrived from www.nea.gov.vn.

Nguyen, Q. T., & Maclaren, V. W. (2005). Community concerns about landfills: A case study of Hanoi, Vietnam. *Journal of Environmental Planning and Management*, *48*(6), 809–831.

Nguyen, Q. T., & Maclaren, V. W. (2007). The evolution of community concern about landfills in Vietnam. *International Development Planning Review*, *29*(4), 413–432.

Okeke, C. U., & Armour, A. (2000). Post-landfill siting perceptions of nearby residents: A case study of Halton landfill. *Applied Geography*, *20*, 137–154.

O'Rourke, D. (2001). Community-driven regulation: Toward an improved model of environmental regulation in Vietnam. In P. Evans (Ed.), *Livable Cities? Struggles for Livelihood and Sustainability*. Berkeley: University of California Press, Berkeley.

O'Rourke, D. (2004). Community-driven regulation: Balancing development and the environment in Vietnam. Cambridge, MA: MIT Press.

Ostrom, E. (1990). Analyzing long-enduring, self-organized, and self-governed CPRs, governing the commons. Cambridge: Cambridge University Press.

Paldam, M. (2000). Social capital: One or many? Definition and measurement. *Journal of Economic Surveys*, *14*(5), 629–653.

Pargal, S., Gilligan, D., & Huq, M. (1999). *Private provision of a public good: Social capital and solid waste management in Dhaka, Bangladesh.* Social Capital Initiative Working Paper No. 16, The World Bank. Retrieved from www.iris.umd.edu/adass/proj/soccappubs.asp

Pargal, S., & Wheeler, D. (1996). Informal regulation of industrial pollution in developing countries: Evidence from Indonesia. *The Journal of Political Economy, 104*(6), 1314–1327.

Petro, N. N. (2001). Creating social capital in Russia: The Novgorod model. *World Development, 29*(2), 229–244.

Phuong, P. T., & Mol, A. P. J. (2004). Communities as informal regulators: New arrangements in industrial pollution control in Vietnam. *Journal of Risk Research, 7*(4), 431–444.

Pretty, J., & Ward, H. (2001). Social capital and the environment. *World Development, 29*(2), 209–227.

Putnam, R. D. (1993). *Making democracy work: Civic traditions in modern Italy*. Princeton, NJ: Princeton University Press.

Roodman, D. M. (1999). Fighting pollution in Vietnam. Washington, DC: Worldwatch Institute.

Rydin, Y., & Pennington, M. (2000). Public participation and local environmental planning: The collective action problem and the potential of social capital. *Local Environment, 5*(2), 153–169.

Stafford, M., Bartley, M., Sacker, A., & Marmot, M. (2003). Measuring the social environment: Social cohesion and material deprivation in English and Scottish neighborhoods. *Environment and Planning A, 35*(8), 1459–1475.

Tang, S.-Y., Tang, C.-P., & Lo, C. W.-H. (2005). Public participation and environmental impact assessment in Mainland China and Taiwan: Political foundations of environmental management. *The Journal of Development Studies, 41*(1), 1–32.

Thanh, T. (2006). Thach Ban: Ca Pho Ra Duong Chan Xe Tai, Xa-Hoi. Retrieved March 17, 2006 from www.vietnamnet.com.vn

Timothy, J. D. (1999). Participatory planning: A view of tourism in Indonesia. *Annals of Tourism Research, 36*(2), 371–391.

To, L. (1993). *Special relationships between traditional Viet villages, the traditional village in Vietnam.* Hanoi: The Gioi Publishers.

Tosun, C. (2000). Limits to community participation in the tourism development process in developing countries. *Tourism Development, 21*(6), 613–633.

Walsh, E. J., & Warland, R. H. (1983). Social movement involvement in the wake of a nuclear accident: Activists and free riders in the TMI area. *American Sociological Review, 48*(6), 764–780.

Reassessing the Voluntary Facility-Siting Process for a Hazardous Waste Facility in Alberta, Canada 15 Years Later

Jamie Baxter

INTRODUCTION

There is a deepening worldwide waste-disposal crisis in the sense that several places continue to struggle to find new places to dispose of wastes that cannot be recycled or reused. Though new approaches to thinking about "waste" are taking hold (e.g., cradle-to-grave product design and responsibility), in the short term at least, we will continue to need such waste sites (Pushchak & Rocha, 1998). A key problem to finding such sites is opposition from local residents, and increasingly, from nonlocals. Traditionally, many jurisdictions relied on hierarchical and technically and professionally driven processes, whereby governments hired experts and together decided on a site that the government would announce to oftentimes unsuspecting communities. At its worst this was the so-called decide-announce-defend approach (Kunreuther et al., 1993). More recently such processes have been modified to include significantly more community participation, but at the core many have remained hierarchical and technical, prone to the same debilitating community opposition as decide-announce-defend.

It is in this context that voluntary facility siting grew to be one of the preferred approaches for finding sites for particularly contentious point-source technological environmental hazards, like hazardous waste and nuclear waste facilities (Kunreuther, Fitzgerald, & Aarts, 1993; Rabe, 1994; Stencel & Lee, 2004). The value of such an approach is that it tends to overcome the problem of "host community" opposition by focusing more attention on benefits (e.g., jobs), rather than just the risk of contamination and other potential negative impacts. Further, there is an element of consent and, ostensibly, control, whereby potential host communities vote in a local referendum (plebiscite) to decide their own fate (Linnerooth-Bayer & Fitzgerald, 1996; Kuhn & Ballard, 1998). Yet, because of its relatively newfound status, critical attention from academics and practitioners alike has focussed mainly on the degree to which this process overcomes the shortcomings of traditional siting, rather than on the degree to which such a process satisfies broader principles of just/equitable siting (Rabe, 1994; Stencel & Lee, 2004; Kuhn & Ballard, 1998). For example there are few studies that assess voluntary siting in philosophical/theoretical terms (Linnerooth-Bayer & Fitzgerald, 1996) or from the point of view of locals who have to live with the results of a voluntary siting process (Gowda & Easterling, 2000).

In particular, useful insights may be gained by looking at the processes in practice, the communities that "host" an actual facility, and other communities affected by the same facility. As Puschak and Rocha (1998) point out, we should be cautious about the long-term success of voluntary siting, since there has been a relative "lack of systematic analysis of voluntary siting outcomes and empirical evidence that voluntary siting outcomes have been more successful than other approaches" (p. 27; see also Rabe, 1994). Such evaluations require engaging with residents and other stakeholders, or as Gowda and Easterling assert, "This task will not be solved within the minds of policy analysts, but requires a journey into the cultures, belief systems, and experiences of the various groups that have a stake" (2000, p. 920). This paper addresses these issues by reporting the results from research related to the Swan Hills (hazardous waste) Treatment Center (SHTC)[1] in Alberta, Canada. The next section briefly reviews principles of just/equitable siting in relation to voluntary siting, particularly as they pertain to the ostensibly successful SHTC process. This is followed by a review of the principles in the context of the resident's views of facility siting, intercommunity relations, and the facility itself.

VOLUNTARY SITING

The Promise of Voluntary Siting

A key advantage of voluntary siting over traditional approaches relates to consent and equity—yet these are also the areas of most criticism. It is

---

[1] Because the facility has changed ownership numerous times it has had different names. It is also known in the siting literature and elsewhere as the Alberta Special (hazardous) Waste Treatment Facility (ASWTF) or the Alberta Special Waste Treatment Center (ASWTC) at Swan Hills.

important to put the voluntary process in the context of how it improves on the decide-announce-defend approach. The voluntary process is meant to address concerns about fairness (equity) raised by nonvoluntary approaches, by virtue of allowing host communities to show their (majority) willingness to host a site, through a local plebiscite (Rabe, 1994; Gowda & Easterling, 2000; Castle & Munton, 1996). The latter is particularly important since it presumably allows a sometimes "silent" majority to approve a facility rather than allow a vocal minority to prevent the facility (Kuhn & Ballard, 1998). Further, that majority decision is supposed to be achieved only after a thorough process of information gathering and interpretation related to things like facility need, site suitability, and safety. Ideally these assessments happen through experts hired by the community with funds provided to the community by the proponent. The process is based on the Pareto-optimal principle that since the host community is presumably better off (e.g., new jobs) and the rest of society is better off (i.e., they have a new place to put their waste), voluntary siting is better than not building a new facility (Linnerooth-Bayer & Fitzgerald, 1996). This is an important equity consideration in the context of an era of rapidly aging/filling local sites. That is, this acknowledges that current host communities have taken *their* turn dealing with the hazard; and that the status quo is unfair if it is allowed to continue if other communities avoid taking their own turn. In a similar sense, the process allows these other communities to show their altruistic side by offering to host new facilities. In this sense a facility may actually foster community pride (Petts, 1995; Baxter & Lee, 2004).

- *Canadian Example—Hazardous Waste Treatment Facility at Swan Hills*

Canadian examples of voluntary siting have often been cited as landmarks of how siting can and should be done successfully. In particular, the voluntary process that resulted in the 1987 construction of the hazardous waste treatment facility fifteen kilometers northwest of Swan Hills (pop. 2,500), Alberta is heralded as a noteworthy success (Rabe, 1994; Stencel & Lee, 2004; Kuhn & Ballard, 1998; Castle & Munton, 1996; Kunreuther et al., 1996). For example, Kuhn and Ballard proclaim,

> The Swan Hills integrated hazardous waste facility, located in north central Alberta, stands as a hallmark of siting success. The process began in 1981 and concluded in 1984. The approach used was clearly successful considering what was achieved and at what cost. Where other hazardous

waste programs have floundered, the Province of Alberta sited a facility in three years at a relatively low cost of $Can.5 million. (1998, p. 537)

Nevertheless, cost savings and timely siting should not be the only considerations for declaring siting success. The way the process was carried out has also received considerable praise since it simultaneously addressed overall facility need, alternatives, facility location, and perhaps above all, community participation and assent through a plebiscite. In fact, in a book on the topic, Rabe carefully argues that the process turned the NIMBY syndrome on its head. That is, rather than individuals/communities espousing the need for waste facilities, but at the same time refusing to host such facilities, i.e., NIMBY, the Alberta's hazardous waste-facility was actually fought over by at least two potential hosts. Rabe recounts the dramatic fact that Ryley, a community of 500 and also over 100 km nearer than Swan Hills to the provincial capital of Edmonton, publicly protested the decision by the provincial government to "award" the facility to Swan Hills. That it was considered an "award" has a lot do to do with the fact that a hospital, other improved infrastructure, and 100+ local jobs all eventually accrued to Swan Hills alone. The reason provided for choosing Swan Hills over Ryley, besides some likely political manoeuvres by the former's provincial representative (Rabe, 1994), is that Ryley was closely located to several other towns and only 80 kms from the capital, Edmonton. The choice of Swan Hills helped avoid two equally problematic solutions for following through with a Ryley facility: (a) holding a multi-community plebiscite or (b) dealing with opposition from other local communities that would likely not receive any direct benefits if the existing Ryley-only vote in favour of the facility were upheld. This issue of the scale of community plebiscites is a central theme of this manuscript as it also applies to the Swan Hills case.

Limitations of Voluntary Siting (Philosophy, Experience, and Practice)

- *Philosophy of Justice—Individualist/Market Worldviews vs. Egalitarian Worldviews*

Environmental injustice refers to replication of historical injustices experienced by disadvantaged groups (e.g., racial minorities, low-income groups) in current realms of hazardous facility cleanup and facility siting (Pulido, 1996). By distinction, environmental equity tends to pay less attention to the past and instead focuses on either procedural equity (e.g., public participation in facility siting) or the spatial distribution of hazards relative to disadvantaged groups, regardless of how the pattern emerged—outcome equity. From a philosophical/moral standpoint environmental justice is a multidimensional concept and voluntary siting tends to emphasise certain forms of equity, potentially at the expense of others, and often disregards environmental justice altogether (Pulido, 1996). For example, the typical version of voluntary siting is market driven and, according to cultural theory, there is a tacit assumption that all involved support an individualistic-competitive worldview; that is, it is assumed that individuals and groups should be free to compete for mutually beneficial gains as long as nobody else (i.e., society at large) is any worse off. Yet, others subscribe to an egalitarian worldview that would ensure that the highest consumers of waste-producing goods—the wealthy—should likewise bear the highest responsibility for the waste, for example hosting a waste facility (Linnerooth-Bayer & Fitzgerald, 1996).

From a justice point of view, among the egalitarians' concerns is the fact that the communities that invariably come forward as potential voluntary siting hosts tend to be challenged with multiple disadvantages including low income, high unemployment, and visible minority status. For instance, Gowda and Easterling describe how Native American communities have tended to be the most prominent groups offering to host monitored retrieval storage facilities for nuclear waste in the United States (Gowda & Easterling, 2000). In the words of one Native American representative, "Just because there are two willing partners to do this tango is no reason to hold the dance" (Linnerooth-Bayer & Fitzgerald, 1996, p. 127). Though there are various mechanisms for sorting out the actual compensation to ensure something better than the lowest "bid" is actually paid to the "winners," the egalitarian's concern is that the reason for even bidding in the first place is due to a position of vulnerability and reduced bargaining power (Rawls, 1999). For example, there are other cleaner industries that pose lower direct health risks that could likewise benefit disadvantaged communities. These are recurrent themes in the environmental equity and justice literatures. The justice literature in particular is concerned with the *intentional* targeting of these vulnerable groups, as opposed to simply a coincidence of "neutral" market forces (Pulido, 1996; Mohai & Saha, 2007). It is for these reasons that egalitarian/justice-oriented writers are among the strongest supporters of increased efforts at waste reduction outright, to avoid the need for large-scale waste disposal facilities in the first place (Linnerooth-Bayer & Fitzgerald, 1996; Thompson et al., 1990). For example, for some facilities, dread of catastrophe may be so high that even financially compromised groups may not be willing to host them (Hine et al., 1997).

• *Practice—Informed Consent*

In terms of the practice of voluntary siting, the manner in which informed consent is carried out is critical. While communities typically vote for or against a facility, who gets to vote in the first place has not been critically assessed. For example, what counts as a reasonable distance from a facility to warrant participation in the decision process and negotiation for compensation? Is distance even the key consideration (Baxter & Greenlaw, 2005)? Gowda and Easterling (2000) report that though a single community may vote to host a facility, their neighbours may be excluded from such a plebiscite or voluntarily exclude themselves on ethical grounds. One result is strained intercommunity relations.

• *Practice—Distributive Equity*

As if intercommunity conflicts were not discouraging enough, excluding nearby communities also violates principles of distributive/spatial/geographic equity if compensation is localized to only a single community. The distribution of costs and benefits needs to be considered on a scale commensurate with the scale of the facility, whereby a large-scale facility should require regional scale (e.g., county, state/province) compensation arrangements (Gowda & Easterling, 2000). The latter is further necessitated on the grounds that the management of large facilities might better be left to oversight more diffuse than that of the operator and a small local community alone (Bradshaw, 2003).

• *Subjective Experience—Stigma vs. Pride*

Another objection to the market-driven voluntary facility siting approach is that the focus on compensation crowds out altruistic reasons for hosting—particularly by more affluent communities. Wealthier communities typically do not "need" waste facilities, but little attention gets placed on convincing them to host facilities if siting is cast as an economic choice rather than a choice to "do ones' part for the environment." Some explain it to be a result of the NIMBY phenomenon (Rabe, 1994), while others are concerned about compensation being misconstrued as bribery (Kunreuther, 1995) or exploitation (Linnerooth-Bayer & Fitzgerald, 1996). A related concern, though, is that the focus on economics and not on altruism creates as a by-product the impression that these facilities remain sources of contamination. Whereas communities should be applauded for being willing hosts, they are, potentially, chided for being environmental dupes (Slovic et al., 1994).

SURVEY FINDINGS

This section reports findings from a survey conducted in the communities of Swan Hills, Kinuso, and Fort Assiniboine as they pertain to the aspects of voluntary siting discussed above. It is important to realize that the overall project was not meant to study voluntary siting per se, it was a multimethod (interviews, survey) study aimed at understanding the social construction of environmental hazard risk in residents' everyday lives. The results here are mainly from a telephone survey conducted with a stratified random sample of 453 residents in 2002, with an overall response rate of 69 percent. The methodology for this study and related studies on the same communities is detailed elsewhere (Baxter & Lee, 2004). Likewise the site history and community characteristics are described elsewhere, but for context here, Swan Hills (pop. 2,500) is closest to the facility (15 kilometers) and can be characterized as a resource town (oil/gas, forestry, waste); Fort Assiniboine is 70 kilometers to the south and is predominantly an agricultural community; while Kinuso 70 kilometers to the north is surrounded by First Nations reserves, comprised largely of First Nations people, and sustained largely by tourism, agriculture, and hunting/trapping.

Philosophy of Justice

By some measures, the SHTC process is a success since, at first glance, it appears to satisfy the problem of environmental justice. We did not approach the problem by asking residents their philosophy of justice according to cultural theory. Instead, it is potentially useful to look at the initial "bargaining position" of Swan Hills residents in the early 1980s, the time when the siting process began. Graph 1 shows that around the time the facility-siting process was happening (1981) the median household income was already above the provincial median ($29,000 vs. $24,500). If Swan Hills was at an economic disadvantage, it was not a dramatic one relative to the province. Further, compared to Ft. Assiniboine and Kinuso, Swan Hills has expectedly remained relatively wealthy for the 20 years since the process happened and the 15 years since the facility became operational. From the point of view of environmental justice, Swan Hills was not a victim at any time.

Yet there can be at least two claims made against this facility from an egalitarian point of view. First, though the facility is located near a relatively wealthy community, as it is a small community, Swan Hills can hardly be considered a major producer of hazardous waste. This raises questions about sustaining motivations for waste reduction within larger centres in the province that *are* the major waste producers. Second, and more importantly, there has been a legacy of injustices against First Nations communities in Alberta. The siting process did not adequately acknowledge the unique way of life of the First Nations communities along Lesser Slave Lake near Kinuso (Baxter & Greenlaw, 2005; Gibson & Froese, 2004). Not only were the Lesser Slave Bands not allowed to vote in the plebiscite to decide the facility's original fate, they were not beneficiaries any of the initial compensation. For example, though the facility operators did pay for health studies in the area through Provincial Court–arranged pollution fines (Baxter & Lee, 2004), no First Nations member has worked at the facility and no health care facilities have been built in their area as a direct result of the siting process. That is, no lasting compensation has been felt.

Figure 1. Canadian Census: Median Household Income for Selected Alberta Communities, 1981–2001

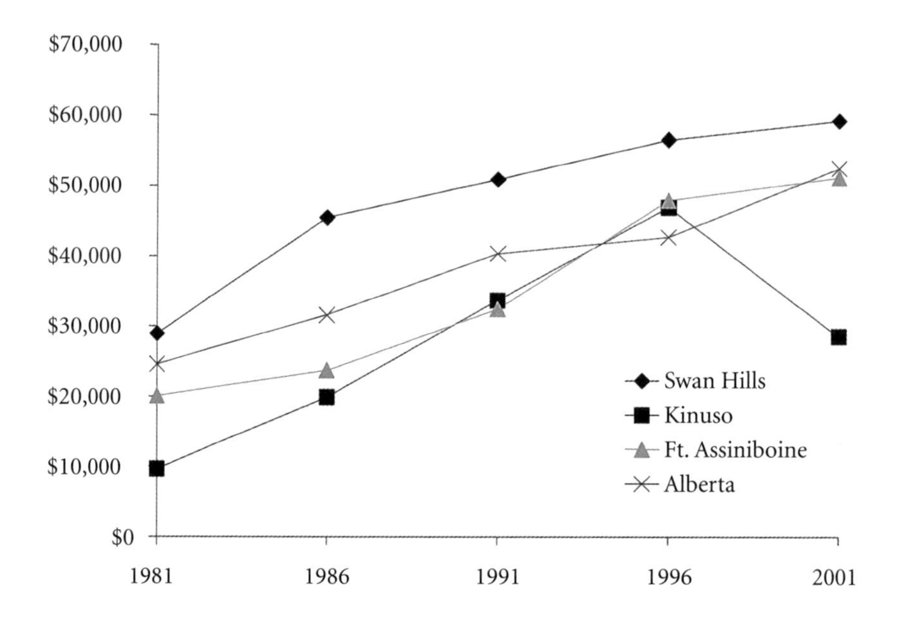

Informed Consent

Informed consent is potentially divisive, even in sparsely populated areas like rural Alberta. Though the First Nations communities were not actually studied directly by our group (Baxter & Greenlaw, 2005, see Gibson & Froese, 2004), the reactions in the towns of Kinuso and Fort Assiniboine provide further insight into the issue of fairness, informed consent, and voluntary siting. Given that our research happened between 1998 and 2002, any specifics about perceptions at the time of the voluntary siting process would be subject to recall biases. Thus, we focused on two key questions in our survey: "If you had the opportunity to vote today on whether the facility should be exactly where it is today, how would you vote?"; and "How fair do you feel the process was that put the Alberta Special Waste Treatment Facility near Swan Hills?" Figure 2 shows that of decided voters, while those in favour in Swan Hills (79 percent) is exactly the same as the actual result when the facility was first sited (Rabe, 1994), if all three towns were allowed to vote in a plebiscite in 2002, those in favour (49 percent) and those against (51 percent) were almost even. Notably, the largely First Nations community of Kinuso opposed (90 percent) the facility by a large margin, while the predominantly agricultural community of Fort Assiniboine likewise opposed it (54 percent).

Perceived fairness shows a similar pattern with the striking difference that the number of undecided residents in Swan Hills is the highest of all the results. Though the large majority of *decided* voters in Swan Hills (90 percent), and a smaller majority overall (56 percent) felt the process was fair, a majority in Kinuso (75 percent) and Fort Assiniboine (52 percent) did not. What is most interesting is that 36 percent from Swan Hills would not express their opinion—a statistically significant difference from the number of undecided voters in the other two towns. Given there is a general reluctance of residents to say anything bad about Swan Hills (Baxter & Lee, 2004), it may be that these people have concerns about the fairness of the process that they would rather not openly express, despite the anonymity of the survey. Further, this fairness measure is a strong predictor of facility-related safety concern. In fact, fairness was typically the most important predictor of facility-related safety concern (all towns, Kinuso, Ft. Assiniboine), more than perceived economic or social benefits, and an effect of similar size to various types of "trusted" information sources (Baxter, 2009).

Distributive Equity

There is a mismatch between (involuntary) risk and benefits on a regional scale. In terms of regional benefits, Figure 1 shows that Swan Hills has retained its higher-than-provincial median income since before the facility was located nearby, and since the over 100 operational jobs came to the town from the facility in 1987. By comparison, the towns of Kinuso and Ft. Assiniboine do not seem to be any worse off relative to Swan Hills over the same time period—with the exception of Kinuso in 2001. Thus, on the basis of median

Figure 2. Hypothetical Vote for Existing SHTC "Today"

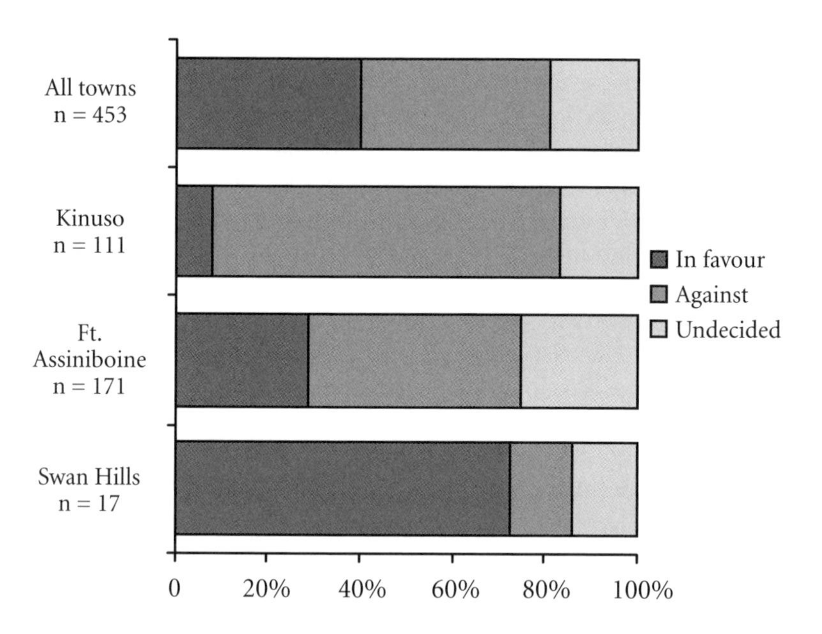

Figure 3. Perceived Fairness of Voluntary Facility Siting for SHTC

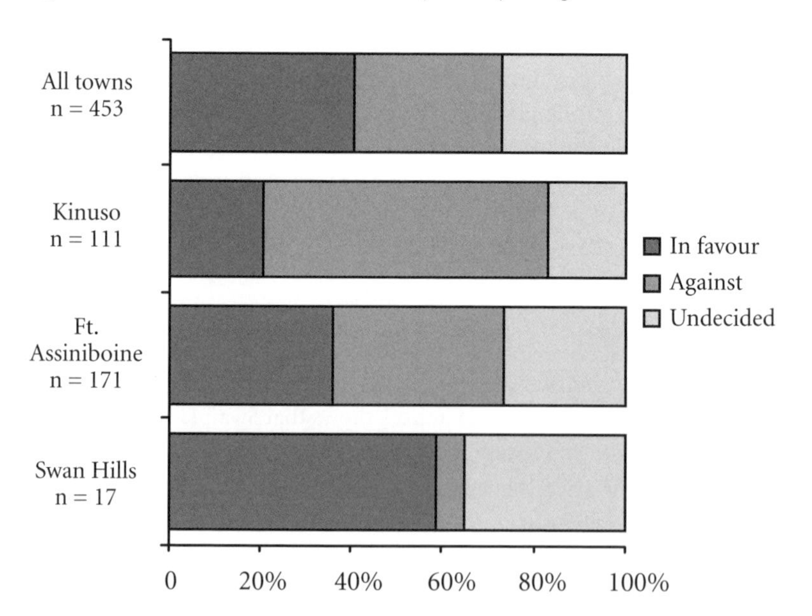

income one might dismiss claims of distributive *in*equity, on the basis that the relative gaps have remained fairly constant. Nevertheless, the amenities—including a 24-bed hospital—went to Swan Hills alone as part of the negotiated compensation package.

In terms of negative impacts, the First Nations communities surrounding Kinuso have successfully argued that the potential negative impacts from the facility extend far beyond the 15-kilometer radius that encompasses Swan Hills (Gibson & Froese, 2004). This idea was reinforced in 1996 when a major leak of PCBs, dioxins, and furans set in motion a series of hunting and fishing bans that affected a large portion of the region surrounding the facility that extended to Fort Assiniboine and Kinuso (Jardine, 2003). In effect this was the realisation of the fears expressed in at least two major environmental assessment processes related to both facility siting (1984) and the eventual expansion of the facility (1995). It is questionable, though, whether Kinuso or Fort Assiniboine residents would have accepted compensation for fear that it might be interpreted as a means to keep them quiet about future facility malfeasance (Kunreuther, 1995). This seems likely given the general sentiment against the facility in Kinuso especially. Compensation, when provided post hoc, can also create problems of intercommunity conflict as demonstrated in one Swan Hills resident's comments about the legitimacy of First Nations complaints:

> Resentment came in the community when they read media articles saying that the "First Nations or Aboriginal people [were] living near the plant."... well they don't. The closest ones live at Kinuso, the Swan River band, and that's about 70 km from the plant. The other native people involved in this action live 100 and 120 km away from the plant. They do visit occasionally to hunt moose and that's it. (David, does not work at SHTC)

As we argue elsewhere, there needs to be mutual respect for different ways of life, whereby hunting moose may be central to one way of life, and merely an indulgence to another (Baxter & Lee, 2004).

Pride vs. Stigma

The majority of Swan Hills residents want the facility, and are indeed proud of the part they are playing in dealing with the province's hazardous waste. In the words of one resident, "I feel we are doing a favor to the rest of the world, like somebody has to look after this and a lot of other places were too afraid to" (Anne, does not work at SHTC). Further, when presented with the following two statements in the survey: "Outsiders are saying bad things about Swan Hills without the facts" and "Swan Hills does not get enough outside credit for hosting the facility," an almost-unanimous 98 percent and 93 percent of the 171 respondents agreed. Yet, this combination of pride and the stigma from "outsiders" has created a unique situation whereby the facility operators, and indeed the community are motivated to keep negative reports about the facility (and town) secret from outsiders. For example, the leak in 1996 was not reported to the provincial authorities until three days after it was known. Further, in a related paper we argue that this situation continues to have the potential to frustrate appropriate facility monitoring (Baxter & Lee, 2004), a situation that might not exist under a more regionalized management scheme (Bradshaw, 2003). In the words of one resident who has latent concerns (Baxter & Lee, 2004) about the facility, but who is nevertheless frustrated with the bad press they repeatedly get,

> The media, I have no respect for [them]. They come up here with full intentions of taking a perfectly good thing and then they turn around and turn it into something ugly, and I have no respect for them whatsoever. They don't get their facts right, they're all wrong. They don't even get their names right. So I have no faith in them and they're the ones feeding the public the wrong information and if they're going to do it properly, sure you've got your negative side, but don't forget about the positive side, you know. You never, ever see them write about the positive side. (Dagmar, does not work at SHTC)

It is people like Dagmar who are torn between a commitment to their community and a desire to inform "outsiders" about facility-related problems as necessary.

DISCUSSION AND CONCLUSIONS

The SHTC voluntary siting process may *not* be as successful as some have argued (Rabe, 1994; Stencel & Lee, 2004; Kuhn & Ballard, 1998; Castle & Munton, 1996). Certainly it was successful in the sense that a facility has been treating the province's hazardous waste for 20 years and the town of Swan Hills was not financially, racially, or historically disadvantaged when they volunteered to be hosts. Indeed they were and are a relatively wealthy community who are arguably doing their part for the greater good.

Though the people of Swan Hills are generally satisfied with the local hazardous waste treatment facility (SHTC), the towns of Fort Assiniboine and, especially, Kinuso are much less content. The reason for this situation can be traced back to the voluntary siting process itself since it involved both an inadequate process of informed consent and unfair initial compensation.[2] It is doubtful, though, that *any* compensation would satisfy these latter two communities given their general dislike of the facility. A key predictor of this dislike can be traced to the perceived fairness of the original siting process that excluded all but the residents of Swan Hills despite the fact that a facility of the SHTC's magnitude has the potential for negative impacts on a regional scale.

Thus, the definition of "community" is central to the voluntary siting process. Indeed, there may be some useful direction provided in the environmental assessment literature as it relates to directly affected parties (Lawrence, 2007). Of paramount importance though is that the scale of "community" affected should be on the same level as the scale of potential negative impacts. There is ample evidence here to question the legitimacy of the single-community–single-facility negotiated settlement model. Though involving multiple communities complicates the process considerably, there are at least two key benefits to such an approach. First, it goes a long way towards satisfying the principles of procedural and distributive equity. Second, it could potentially lead to a greater emphasis on the altruistic roles played by the communities, rather than incite intercommunity conflict and put single-town residents in a mindset of being closed to outsiders.

Whether a process involving a scaling up of "community" addresses widespread concerns about justice—vis à vis an initial disadvantaged bargaining position for certain groups—is uncertain. Certainly this will depend on the types of communities involved. Due attention needs to be paid to the historical legacy of any injustices to any group that becomes involved in negotiations. Further, the process should confront head-on alternative means for rectifying injustices, beyond just a potentially risky facility.

There is an ongoing urgency to debate issues surrounding voluntary siting, a dialogue that has waned in recent years in the environmental management literature. Voluntary siting remains one of the preferred ways to locate sites for disposing hazardous waste and nuclear wastes in Canada. For example, voluntary siting is the means by which Canada's Nuclear Waste Management Organization (www.NWMO.ca) will find a deep geologic repository for Canada's high-level nuclear wastes. Whether or not the manner in which the NWMO defines potential community hosts has a bearing on the "success" of the process will be telling.

---

[2] Compensation agreements have been reached between the operator and First Nations communities, but these had been won in environmental appeals rather than during the siting process.

REFERENCES

Baxter, J. (2009). Risk perception: A quantitative investigation of the insider/outsider dimension of cultural theory and place. *Journal of Risk Research*, *12*, 771–791.

Baxter, J., & Greenlaw, K. (2005). Revisiting cultural theory of risk: Explaining perceptions of technological environmental hazards using comparative analysis. *Canadian Geographer*, *49*, 61–80.

Baxter, J., & Lee, D. (2004). Explaining the maintenance of low concern near a hazardous waste treatment facility. *Journal of Risk Research*, *6*, 705–729.

Bradshaw, B. (2003). Questioning the credibility and capacity of community-based resource management. *The Canadian Geographer*, *47*, 137–150.

Castle, G., & Munton, D. (1996). Voluntary siting of hazardous waste facilities in Western Canada. In D. Munton (Ed.), *Hazardous waste siting and democratic choice*. Washington, DC: Georgetown University Press.

Gibson, G., & Froese, K. (2004). *Hazardous waste: Disrupted lives (First Nation Perspectives on the Alberta Special Waste Treatment Centre)*, Environmental Health Sciences, Edmonton, University of Alberta.

Gowda, R., & Easterling, D. (2000). Voluntary siting and equity: The MRS experience in Native America. *Risk Analysis*, *20*, 917–929.

Hine, D., Summers, C., Prystupa, M., & McKenzie-Richer, A. (1997). Public opposition to a proposed nuclear waste repository in Canada: An investigation of cultural and economic effects. *Risk Analysis*, *17*, 293–302.

Jardine, C. (2003). Development of a public participation and communication protocol for establishing fish consumption advisories. *Risk Analysis*, *23*, 461–471.

Kuhn, R., & Ballard, K. (1998). Canadian innovations in siting hazardous waste management facilities. *Environmental Management*, *22*, 533–545.

Kunreuther, H. (1995). Voluntary siting of noxious facilities: The role of compensation. In O. Renn, T. Webler, & P. Wiedemann (Eds.), *Fairness and competence in citizen participation: Evaluating models for environmental discourse*. Dordrecht, Netherlands: Kluwer Academic, 1995.

Kunreuther, H., Fitzgerald, K., & Aarts, T. (1993). Siting noxious facilities: A test of the facility siting credo. *Risk Analysis*, *13*, 301–318.

Kunreuther, H., Slovic, P., & MacGregor, D. (1996). Risk perception and trust: Challenges for facility siting. *Risk, Health, Safety and Environment*, *7*, 109–118.

Lawrence, D. (2007). Impact significance determination: Designing an approach. *Environmental Impact Assessment Review*, *27*, 730–754.

Linnerooth-Bayer, J., & Fitzgerald, K. (1996). Conflicting views on fair siting processes: Evidence from Austria and the U.S. *Risk, Health, Safety and Environment*, *Spring*, 119–134.

Mohai, P., & Saha, R. (2007). Racial inequality in the distribution of hazardous waste: A national-level reassessment. *Social Problems*, *54*, 343–370.

Petts, J. (1995). Waste management strategy development: A case study of community involvement and consensus-building in Hampshire. *Journal of Environmental Planning & Management*, *38*, 519–536.

Pulido, L. (1996). A critical review of the methodology of environmental racism research. *Antipode, 28*, 142–159.

Pushchak, R., & Rocha, C. (1998). Failing to site hazardous waste facilities voluntarily: Implications for the production of sustainable goods. *Journal of Environmental Planning and Management, 41*, 25–43.

Rabe, B. (1994). *Beyond NIMBY: Hazardous waste siting in Canada and the United States.* Washington, DC: Brookings Institution.

Rawls, J. (1999). *A theory of justice.* London: Oxford University Press.

Slovic, P., Flynn, J., & Gregory, R. (1994). Stigma happens: Social problems in the siting of nuclear waste facilities. *Risk Analysis, 14*, 773–777.

Stencel, J., & Lee, K. (2004). The voluntary siting process, a case study in New Jersey. *Health Physics, 86*, S57–63.

Thompson, M., Ellis, R., & Wildavsky, A. (1990). *Cultural theory.* Boulder, CO.: Westview Press.

11

Structural Model of Risk Perception on Landfill Site for Municipal Solid Waste

Kaoru Ishizaka, Yasuhiro Matsui, and Masaru Tanaka

INTRODUCTION

In Japan, municipal solid waste generates more than 50 million tons per year, 80 percent of which is incinerated, and 7.3 million tons of ash and non-combustible wastes which is disposed of in landfill sites with leachate control. A landfill site is an essential disposal facility for waste management; however, it is associated with uncertain risks and public protests. Just like other regions in the world, in Japan, waste treatment and disposal facilities are not irrelevant to NIMBY syndromes, especially; shortage of landfill sites is a serious problem for local governments. In order to improve the public acceptability for the landfills, practical use of risk communication is necessary. This calls for analysis concerning the psychological aspects of the structure of risk perception.

Studies of risk perception have been conducted by many researchers, and as a result, it is widely agreed that trust is a key factor influencing people's perceptions of risk (e.g., Slovic, 1993, 1997; Kasperson, 1992). In Japan, in a case study of landfill sites for high-level radioactive waste, Tanaka (1998) stated that the main factors of public acceptance were "concern and care," "information disclosure," "communication with citizens," "compensation in case of accident," and "risk management ability."

This study aims to analyse the factors relevant to the acceptance and risk perception of landfill sites for municipal solid waste, focusing on public trust, and to construct the structural model to understand the relationship among these factors generally.

METHODOLOGY

Data

Our case study was conducted in Okayama prefecture, in the western part of Japan. Yoshinaga City, Kurashiki City, and Okayama City were selected as research areas (Figure 1). Table 1 shows basic data, and Table 2 shows the historical background of waste management of each city. The city of Okayama is the central area of Okayama prefecture, characterised by commercial areas around the central station. The city of Kurashiki is an industrial area, characterised by heavy industry and textiles. The city of Yoshinaga is a rural area, characterised by a dependence on agriculture. And in Yoshinaga City, there was public conflict concerning the siting plan for a landfill site for industrial waste from 1994 to 1998.

One thousand residents living in three cities were selected by systematic random sampling from the telephone directory, and surveyed by the mailing method. The number of valid responses was 423. Of those, the sex distribution was male 70.4 percent; female 28.3 percent. Age distribution was as follows: 0–29, 1.4 percent; 30–39, 3.3 percent; 40–49, 13.0 percent; 50–59, 25.5 percent; 60–69, 30.0 percent; and, over 70, 23.6 percent.* Sex and age distribution showed no significant differences among the three cities.

Figure 1. Location of Research Area

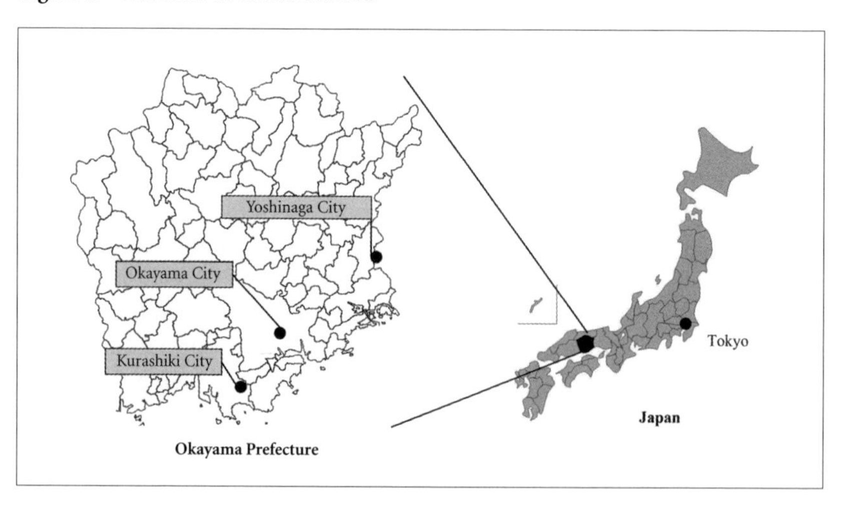

* Percentages may not add up precisely due to rounding.

Table 1. Basic Data of Research Area

Research area	Population*	Number of households	Size of household	Necessary sample number
Okayama City	626,534	244,010	2.57	96
Yoshinaga City	5,288	1,690	3.13	91
Kurashiki City	430,239	152,510	2.82	96

* National population census 2000.

Table 2. Historical Background of Waste Management in Research Area

Yoshinaga

In 1994, a joint public-private venture submitted a development plan for an industrial waste landfill site to the Yoshinaga city government. Then, a residents' group started a campaign against the proposed plan, and that movement spread citywide. On September 16, a referendum was held about whether to accept the landfill site; about 90 percent of voters rejected it. As a result, the government of Okayama Prefecture decided to take measures to refuse a construction license for the landfill site. The developer appealed the decision of the Okayama Prefecture to the national government, but the Ministry of Health and Welfare upheld the decision of Okayama prefecture's government.

Kurashiki

In 2004, JFE, a steel company, had build an incineration plant, which treated simultaneously municipal and industrial wastes. The JFE's new system was the first of its kind in Japan. This plan has received no major protest by residents.

Okayama

In 2001, an incineration plant was rebuilt in the area, but this work has received no major protest by residents.

Questionnaire Items

Table 3 shows the questionnaire. The acceptance of the landfill site was measured by two questions: "Can you accept the siting of a municipal waste landfill near your residence?" and "Can you accept the waste generated in another area?" Risk perception was measured by two questions asking about the possibility of an accidental leak in a controlled landfill site and the seriousness of environmental pollution caused by the accident. To measure the factors relevant to acceptance and risk perception of the landfill site, we designed the question addressing this issue on the areas of "Necessity," "Benefit," "Trust in technology and standards," "Trust in local government," and "Knowledge of chemical substances." Additionally, to understand the risk perception level of a waste-management facility, the questionnaire included a perception of the necessity and safety of several risk factors (e.g., nuclear power plant, tobacco use). Questions were answered on a seven-point Likert scale.

Table 3. Questionnaire

Acceptance	Q302	Can you accept the siting of a municipal waste landfill site near your residence?
	Q305	Can you accept the waste generated in another area?
Risk perception	Q601	Do you think the possibility of an accidental leak in the landfill site is high?
	Q602	Do you think it seems more likely that the serious pollution of soil and ground water will be caused by the breakage of the landfill liner?
Necessity	Q210	Do you think the landfill site is an essential facility?
Benefit	Q708	Do you want to use welfare provisions (e.g., a pool) attached to the waste treatment facility?
Trust in technology and standards	Q603	Do you think that landfill construction only needs to meet national standards?
	Q604	Do you think that the landfill can be controlled safely if current technology is applied to the landfill?
Trust in local government	Q701R	Do you fear whether you will get compensation if the waste treatment facility does cause pollution?
	Q702R	Do you fear the city may conceal information intentionally if an accident occurs at the waste treatment facility?
	Q703	Do you think the city made an effort to disclose information properly?
	Q705	Do you think the city took into consideration the views (opinions) of residents with regard to the management of the waste-treatment facility?
	Q706	Can you leave the management of the waste treatment facility to the city administration without worry?

Knowledge of chemical substances	Q901	Do you think chemical substances can be divided in two categories: hazardous and safe?
	Q902	Do you think the chemical substances' risks can be reduced to zero?
	Q903	Do you think the hazard of chemical substances is well-understood scientifically?
	Q904	Do you think you are safe if you take in carcinogens below the regulated dose?
Comparison with another risk factor (necessity and safety)		Nuclear power plant, automobile factory, tobacco, pesticide, electromagnetic wave of cell phone, exhaust gas from diesel automobile, food additive, genetically-modified food
Individual attribute		age • sex • civil status • occupation • family size • distance to waste treatment site

Statistical Analysis

At first, to remove any biased sample, we compared the three-city data and individual attributes by nonparametric multiple comparison, and selected data for modelling analysis. Second, we validated the construction of the latent variables using factor analysis. Third, we built and tested the theoretical model with latent variables using the structural equation model.

Comparison of the Three-City Data

Figure 1 shows a comparison of the perceptions of "Necessity" and "Safety" of a waste management facility and other risk factors. The perceptions of necessity of waste-management facilities were highest compared to other risk factors in the cities of Okayama and Kurashiki; meanwhile, the necessity of an automobile factory was highest in Yoshinaga. Perceptions of the safety of industrial waste-management facilities were in the same range as nuclear power plants in Okayama and Kurashiki. On other hand, in Yoshinaga, perceptions of safety of industrial waste facilities were lower than those of nuclear power plants. Overall, perceptions of safety of municipal waste-management facilities were higher than industrial waste, and an incinerator rated higher than a landfill site.

We examined the statistical differences of the three cities by one-way analysis of variance and nonparametric multiple comparison ($p < 0.05$). As a result, Yoshinaga city data shows a significant difference compared to the other cities on several questionnaire items (Q302, Q305, Q601, Q602, Q603, Q604, Q904, Q701, Q702, Q703, Q705, Q706). The mean of these items showed that Yoshinaga citizens have a negative image of waste management (facility, technology, and standards) and local government. This study aims to develop a general model of acceptance and risk perception, thus, we removed Yoshinaga city data from the modelling analysis.

Figure 2. Perceptions of Necessity and Safety Risk Factors

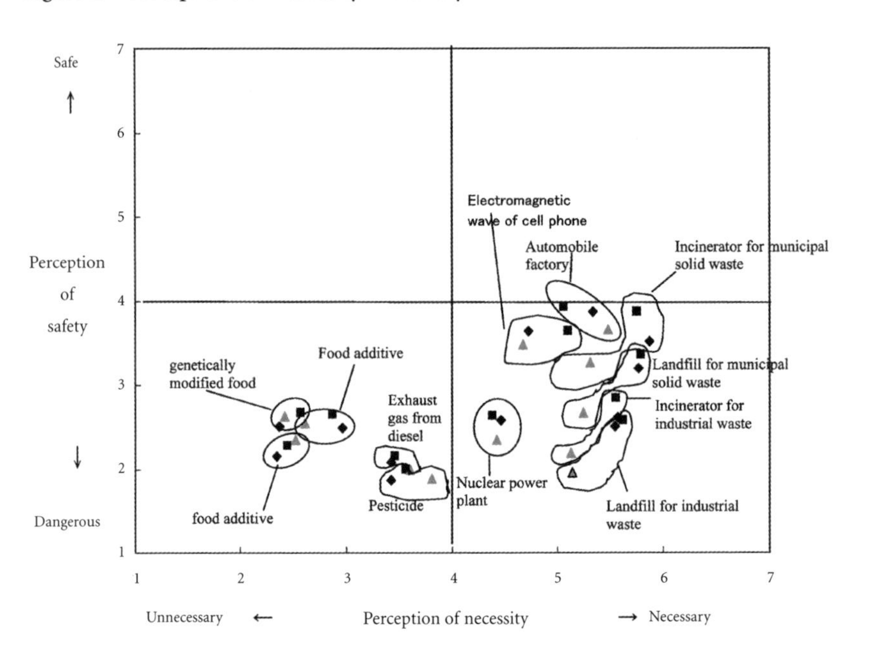

Validation of Latent Variables by Factor Analysis

To clarify the relevant variables of acceptance and risk perception, we examined the correlation between acceptance or risk perception and other questionnaire items. As a result, "Necessity" (Q210), "Benefit" (Q708), and "Knowledge of chemical substances" (Q901–904) show no significant correlation with acceptance and risk perception, so we left these items out of the factor analysis.

Using the remaining items, we undertook explanatory factor analysis to fix the latent variables for structural equation modelling, employed principal factor method with promax rotation. Table 4 shows five factors extracted by factor analysis. Factor 1 was named "Acceptance," Factor 2 as "Risk perception," Factor 3 as "Trust in technology and standards," Factor 4 as "Trust in response to accidents," Factor 5 as "Trust in sincerity to citizens." On the questionnaire design stage, Q904 was set up as "Knowledge of chemical substances," but it loaded to "Trust in technology and standards." And "Trust in local government" was divided into "Trust in response to accidents" and "Trust in sincerity to citizens." Using these five factors as latent variables, we performed structural equation modelling.

Table 4. Factor Analysis for Identification of Latent Variables

	Rotated component matrix Component				
	1	2	3	4	5
Q302	**0.840**	7.550E–02	–5.856E–02	0.106	–1.986E–02
Q305	**0.920**	–5.744E–02	5.873E–02	–9.133E–02	1.888E–02
Q601	2.556E–02	**0.910**	–6.029E–03	–3.031E–02	–1.669E–02
Q602	–1.730E–02	**0.942**	–1.714E–02	3.096E–02	–1.077E–03
Q603	–1.843E–03	8.597E–02	**0.856**	–2.161E–02	0.109
Q604	2.913E–03	7.198E–03	**0.896**	–2.814E–02	8.211E–03
Q904	1.098E–02	–0.132	**0.628**	8.413E–02	–0.144
Q701	–1.396E–03	3.083E–02	2.714E–02	**0.982**	–0.102
Q702	2.594E–03	–3.885E–02	–9.386E–03	**0.841**	0.157
Q703	2.971E–02	–3.576E–02	–6.790E–03	7.682E–02	**0.846**
Q705	4.245E–02	–3.203E–02	–5.299E–02	–9.302E–02	**0.963**
Q706	–8.214E–02	5.778E–02	4.612E–02	4.029E–02	**0.778**

Extraction Method: Principal factor method
Rotation Method: Promax rotation
Boldface values indicate items loading most heavily on each factor

Structural Equation Modelling with Latent Variables: Structural equation modelling was used to analyse the relationship of "Acceptance," "Risk perception," and other latent variables. At first, we developed the starting model (model 1) using "Acceptance" and "Risk perception" as endogenous variables. It was hypothesised that "Acceptance" depended on "Risk perception," "Trust in technology and standards," "Trust in response to accident," and "Trust in sincerity to citizens"; (2) "Risk perception" depended on "Trust in technology and standards," "Trust in response to accidents," and "Trust in sincerity to citizens"; (3) "Trust in technology and standards," "Trust in response to accidents," and "Trust in sincerity to citizens" were in covariant relationship. Model 1 was tested on Okayama and Kurashiki data, and proved to fit, indicating that Model 1 was a plausible causal model of the observed data (χ^2 = 59.617, df = 4, p = 0.058, RMSEA = 0.034). But some of the paths were insignificant. We left out the path that was at the lowest level of significance, and proved to fit over and over again until all path coefficients became significant. In the process, we left out the path from "Trust in response to accidents" the first time (Model 2), the path from "Trust in local government" to "Acceptance" the second time (Model 3), the path from "Trust in local government" to "Risk perception" the third time (Model 4). Table 5 shows estimated path coefficients and t-value of Model 1 and Model 4. All paths of Model 4 were significant, and based on goodness-of fit indicators (Table 6), so we adopted Model 4 as the final model. Figure 3 shows the path diagram of Model 4 (standardised solution). In this model, (1) "Acceptance" depended on "Risk perception" and "Trust in technology and standards" in comparable level; (2) "Risk perception" depended on "Trust in technology and standards" and "Trust in response to accident" in comparable level, and (3) "Trust in technology and standards," "Trust in response to accidents," and "Trust in sincerity to citizens" were in covariant relationship.

Table 5. Estimates of T-model Parameters

From		To	Model 1			Model 4		
			estimates	t-value	p	estimates	t-value	p
Regression weight								
Risk perception	--->	Acceptance	0.467	4.545	***	0.467	4.655	***
Trust in technology and standards	--->	Acceptance	–0.289	–4.190	***	–0.282	–4.645	0.001
Trust in response to accidents	--->	Acceptance	–0.013	–0.151	0.880	–	–	–
Trust in sincerity to citizens	--->	Acceptance	0.024	0.255	0.799	–	–	–
Trust in technology and standards	--->	Risk perception	–0.121	–2.646	0.008	–0.129	–3.188	***
Trust in response to accidents	--->	Risk perception	–0.177	–3.028	0.002	–0.189	–3.666	***

From		To	Model 1			Model 4		
			estimates	t-value	p	estimates	t-value	p
Trust in sincerity to citizens	--->	Risk perception	–0.027	–0.418	0.676	–	–	–
Acceptance	--->	Q0302	1.000	–0.151		0.745	6.663	***
Acceptance	--->	Q0305	0.745	6.666	***	1		
Risk perception	--->	Q0601	1.000			1		
Risk perception	--->	Q0602	0.742	9.278	***	0.745	9.302	***
Trust in technology and standards	--->	Q0603	1.000			1		
Trust in technology and standards	--->	Q0604	0.969	11.879	***	0.971	11.884	***
Trust in response to accidents	--->	Q0701	1.000	–2.646		1		
Trust in response to accidents	--->	Q0702	1.404	–3.028	***	1.392	9.032	***
Trust in sincerity to citizens	--->	Q0703	1.000	–0.418		1		
Trust in sincerity to citizens	--->	Q0705	0.915	4.545	***	0.888	13.167	***
Trust in sincerity to citizens	--->	Q0706	0.887	–4.190	***	0.916	15.095	***
Trust in technology and standards	--->	Q0904	0.389	0.255	***	0.39	6.578	***
Covariances								
Trust in sincerity	<-->	Trust in technology and standards	0.656	6.064	***	0.655	6.069	***
Trust in response to accidents	<-->	Trust in sincerity	0.499	5.195	***	0.503	5.572	***
Trust in response to accidents	<-->	Trust in technology and standards	0.302	3.547	***	0.305	3.576	***

Table 6. Comparison of Goodness-of-fit Measures for the Four Models

Goodness-of-fit measures	Model 1	Model 2	Model 3	Model 4
χ^2	59.617	59.639	59.683	59.847
Degrees of freedom	44	45	46	47
Probability level	0.058	0.071	0.085	0.099
RMSEA: Root Mean Square Error of Approximation	0.034	0.032	0.031	0.029
GFI: Goodness-of-fit Index	0.971	0.970	0.970	0.970
AGFI: Adjusted Goodness-of-fit Index	0.948	0.949	0.950	0.951
RMR: Root Mean Square Residual	0.064	0.064	0.064	0.064
AIC: Akaike's Information Criterion	127.617	125.639	123.683	121.847

Figure 3. Path diagram of the estimated model (Standardised Solution)

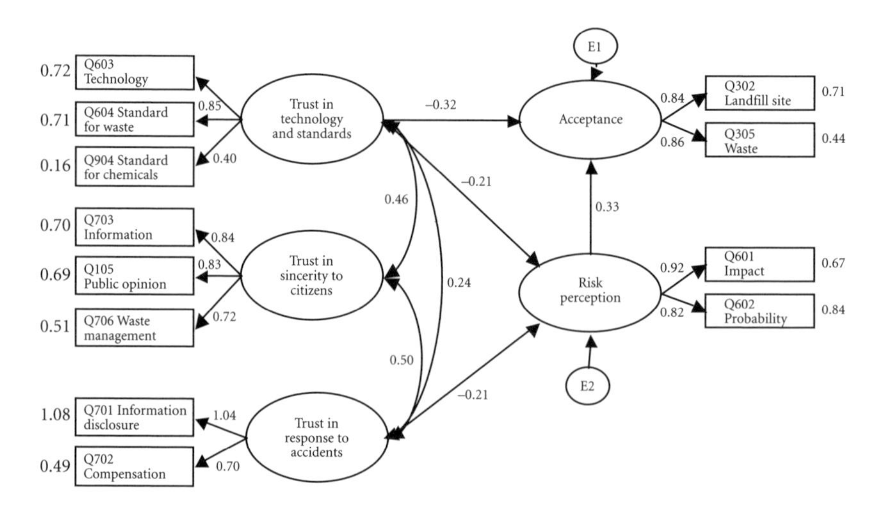

DISCUSSION AND CONCLUSIONS

The results of the questionnaire survey, the mean of the Yoshinaga data show significant differences with the other city data, thus it was revealed that Yoshinaga citizens have a relatively negative conception of waste-management facilities and technology, and to local government. We guess this was because a public conflict occurred in Yoshinaga in 1994–1998. In that case, conflict was caused by development of a landfill site planned by a joint public-private venture, and negative information concerning the landfill site was spread among citizens by an opposition campaign (e.g., "Environmental standards cannot be completely trusted," "Landfill site is very risky.") Because of such a background, it is expected that Yoshinaga citizens perceived the siting as a serious problem.

The result of Structural Equation Modelling indicates that the model has an acceptable fit to the data. All loadings are statistically significant, thus supporting the theoretical basis for assignment of indicators for each latent variable. The model parameters showed "Risk perception" and "Trust in technology and standards" were determinant of citizen's behavioural intentions to accept the landfill site and "Trust in technology and standards" and "Trust in response to accidents" were determinant of citizen's risk perception of landfill sites Wildavsky and Dake (1990) stated that the great struggles over the perceived dangers of technology in our time are essentially about trust and distrust of social institutions. The results of our study affirm the importance of education concerning technology of risk management, and the necessity of monitoring systems in a landfill site. Information disclosure and compensation systems for environmental pollution accidents were also important.

Meanwhile, our model shows "Trust in sincerity to citizens" was in covariant relationship with "Trust in technology and standards" and "Trust in response to accidents." This means, in our model, "Trust in sincerity to citizens" has an indirect effect on risk perception. In this regard, the National Research Council notes that "openness is the surest policy." The results of our study affirm the importance of daily communication between citizens and local government, an open-door policy providing public access to all information, and performance reports about waste management.

In this study, we designed questionnaire items absent a siting process. Although in our earlier survey conducted on actual conflict case, the fairness of siting process was a main factor of acceptance of landfill site. Because fairness of the siting process is expected to have a direct effect on trust in local government, we plan to analyse this aspect in our next survey.

REFERENCES

Fischhoff, B., Lichtenstein, S., Slovic, P., & Keeney, D. (1981). *Acceptable risk.* Cambridge, MA: Cambridge University Press.

Ishizaka, K., & Tanaka, M. (2003). Resolving public conflict in site selection process—A risk communication approach. *Waste Management, 23*(5), 385–396.

Kasperson, R. E., Golding, D., & Tuler, S. (1992). Social distrust as a factor in siting hazardous facilities and communicating risks. *Journal of Social Issues, 48*(4), 161–178.

National Research Council. (1989). *Improving risk communication.* Washington, DC: National Academy Press.

Slovic, P. (1993). Perceived risk, trust, and democracy: A systems perspective. *Risk Analysis, 13*, 675–682.

Slovic, P. (1997). Trust, emotion, sex, politics, and science: Surveying the risk-assessment battlefield. In M. H. Bazerman, D. M. Messick, A. E. Tenbrunsel, & K. A. Wade-Benzoni (Eds.), *Environment, ethics, and behavior: The psychology of environmental valuation and degradation.* San Francisco: The New Lexington Press.

Starr, C. (1969). Social benefit versus technological risk. *Science, 165*, 1232–1238.

Tanaka, Y. (1998). Psychological factors determine public acceptance of landfill site for high-level radioactive waste. *Journal of Japan Risk Research, 10*(45).

Wildavsky, A., & Dake, K. (1990). Theories of risk perception: Who fears what and why? *Daedalus, 119*(4), 41–60.

Compensation in Siting Hazardous Facilities: A Radioactive Waste Repository in Taiwan

Daigee Shaw and Te-hsiu Huang

INTRODUCTION

This paper examines a case where the state-owned Taiwan Power Company (Taipower) attempted to locate a low-level radioactive waste (LLRW) repository in Wu-chiu, a remote island off the coast of Mainland China and under the control of the Taiwan government. The case of Wu-chiu is unique in two respects when it comes to the NIMBY (Not In My BackYard) phenomenon. First, it involves a unique siting process with a generous amount of compensation. Since siting locally unwanted facilities is always a difficult job for governments and developers because of the ubiquitous NIMBY phenomenon, the first solution that comes to our minds is usually compensation. In 1995, after a long period of unsuccessful attempts to site an LLRW repository in Taiwan using the traditional hierarchical approach, Taipower turned to a voluntary siting process with compensation, inviting local townships to voluntarily be candidates in providing sites for the repository to be built. Those townships that voluntarily became candidate sites, once selected, could have been rewarded with various kinds of compensation, depending on the stage of the process involved. Table 1 provides the details of the compensation scheme in relation to the voluntary siting process. In 1996, nine townships entered the first stage and five were deemed to be qualified; however, they all withdrew their applications in 1997 because of the heavy social pressure that resulted from the news media's in-depth coverage of the planned repository. Consequently, Taipower revised its voluntary siting process by adding an expert screening and selection process to the voluntary siting process with its emphasis on the compensation scheme of the voluntary siting process. Then it chose Wu-chiu to be the candidate site.[1]

---

[1] Wu-chiu was one of the five voluntary and qualified candidate communities in 1996.

Table 1. The Compensation Scheme of the Voluntary Siting Process

Phase	Compensation	Condition
First phase	NT$1 million	After signing an agreement with Taipower for further survey and assessment, each township with voluntary but unqualified sites is entitled to compensation of NT$1 million.
Second phase	NT$50 million	After signing an agreement with Taipower for further survey and assessment, each township with voluntary and qualified candidate sites is entitled to compensation of NT$50 million.
Third phase	NT$100 million	Among the candidate sites, Taipower will select some to conduct geological surveys and environmental assessments. For these selected sites, Taipower will offer compensation of NT$100 million.
Fourth phase	NT$3,000 million	Among the assessed sites, Taipower will select one site to construct and operate the repository. Compensation of NT$3,000 million in total will be offered to the local governments surrounding the site.

Second, Wu-chiu's location and social structure is special. It is an isolated island village located between Kinmen and Matsu just off the coast of Mainland China but under the control of the Taiwan government (Figure 1). Before 1949, this small island was a place where fishermen could stay temporarily during the fishing season and there were no permanent residents on the island. However, some fishermen from nearby Mainland fishing villages and their families moved to and stayed on Wu-chiu after 1949 when the Communist Chinese government took control of the Mainland, and the Nationalist Chinese government retreated to Taiwan and kept Wu-chiu, as well as Kinmen and Matsu, two bigger islands along the coast of the Mainland, under its control. Since then, in ways similar to other places following the war, Wu-chiu first experienced a period of population growth. However, in recent years, apart from the army that is stationed there, its population has declined to 30 residents who provide various services for the army because most of the residents have moved to and have remained in Taiwan. Most of them, however, are still registered as residents of Wu-chiu Township because they have strong ties there. There are in total 95 households.

The purpose of this chapter is to examine the residents' perception of and attitudes toward compensation for siting NIMBY facilities using the results of a population-wide survey of the residents registered in Wu-chiu Township that we conducted specifically for this purpose in 2000. The Wu-chiu case is ideal for this purpose because, in addition to the voluntary siting process and the generous compensation offered, it has unique features such as residents who are deeply connected with the place despite being absent.

We first review and identify the factors that affect the residents' perception of and attitudes toward the facility in the next section. We then report on a detailed investigation of the compensation's effects on the Wu-chiu residents' perception of and attitudes toward the facility. After comparing the Wu-chiu case with a similar case of a radioactive nuclear-waste repository in Switzerland, we enrich the model of the compensation cycle developed by Frey et al. (1996) and make it more general. Finally, we offer policy implications in the conclusion.

Figure 1. Wu-chiu's Location (Modified from Google Earth)

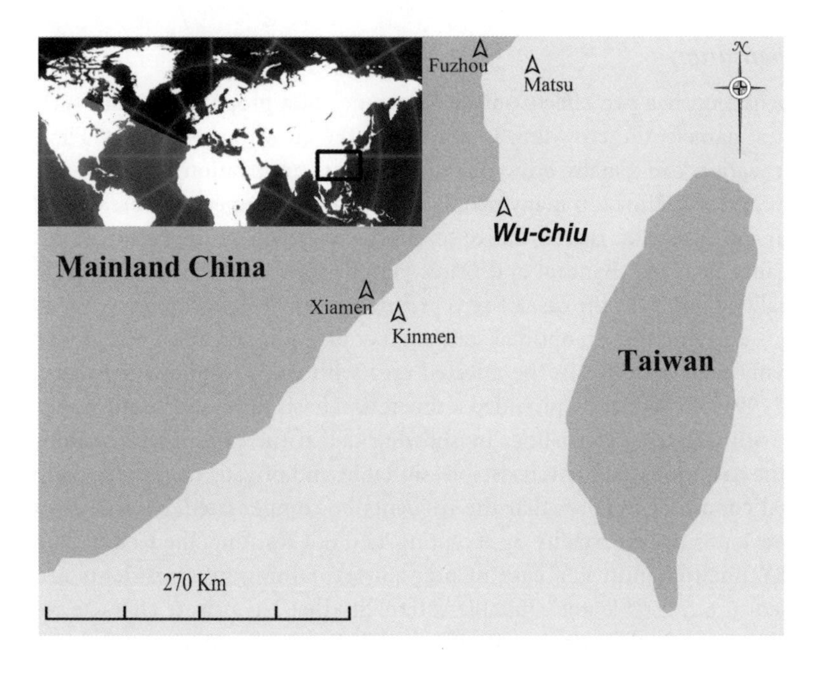

FACTORS THAT AFFECT THE PUBLIC'S PERCEPTIONS OF AND ATTITUDES TOWARD THE FACILITY

Intuitively, the way to reduce the NIMBY phenomenon is to compensate the residents located close to the facility. As long as the level of compensation is sufficiently high, the residents will accept the facility. However, many of the empirical studies argue that the use of compensation in siting highly risky projects is constrained politically by a myriad of procedural, moral, and ethical concerns. Lesbirel (1998) finds that the effect of compensation varies with the locations involved. In some places compensation is effective, while in other places it is not, an outcome that must be due to the differences in each location's characteristics. We therefore review the literature and find that the factors that influence the NIMBY phenomenon and the acceptance of the NIMBY facility include (1) compensation, (2) risk perception, (3) siting procedures, (4) trust in the developers, (5) fairness, (6) how necessary that the facility be built somewhere is perceived to be, (7) social pressures, (8) civic duty, and (9) socioeconomic factors. We discuss these nine factors one by one in what follows:

Compensation

Compensation has two effects on the acceptance of a proposal to build such facilities, namely, the crowding-in effect and the crowding-out effect. First, compensation can usually crowd-in the residents' motivation to accept the facility, and will draw too many people to live in the compensated area. Even though compensation is an intuitive solution, it depends on the nature of the externality involved. Baumol and Oates (1988) showed that if it is a public externality, then, as in the case of zero pricing for public goods, zero compensation is necessary for an optimal solution, because compensation will attract too many people to live in the affected area with compensation. Shaw and Shaw (1991) subsequently provided a theoretical basis for compensating residents with negative externalities by showing that if the externality resulting from the hazardous facility is resistable, shiftable, and private, then the socially optimal condition requires that the residents be compensated, because they generate a positive externality by accepting, and not resisting, the facility. The NIMBY phenomenon is a case in point in explaining how residents are opposed to a resistable and shiftable externality that has private characteristics. Consequently, the compensation in the NIMBY cases justifies itself.

Second, Frey et al. (1996) and Frey and Oberholzer-Gee (1997) observed that compensation has another effect on the acceptance of the building of such facilities, i.e., the crowding-out effect. They probed into cases and found that residents with a strong sense of civic duty will view the compensation as a bribe, which to them represents unethical behaviour, and thus crowds out their motivation to accept the facility for the common good.

Risk Perception

The literature demonstrates that the residents' concerns about risk can be attributed to their sensitivity to the technical, social, and psychological qualities of risks, and there is an important relationship between risk perception and risk acceptance (Slovic, 1993). Kunreuther and Easterling (1992) discover that compensation has a positive effect on the residents if the facility is a safer one, as in the cases of incinerators and landfills, but when it comes to more dangerous facilities such as nuclear power plants, the effects of compensation are insignificant or might even backfire.

Siting Procedures

The residents' acceptance of the siting procedure is an important factor (Frey et al., 1996). The siting procedures today lie between two categories, namely, the hierarchical approach and the voluntary-market approach. These procedures usually cannot work because the best strategy for residents is to stage NIMBY protests under the hierarchical approach, while communities still refuse to cooperate and negotiate siting deals with developers due to other factors such as the crowding-out effect and trust (Kasperson, 2005; Linnerooth-Bayer, 2005; Shaw, 2005).[2]

Trust

The residents' trust in the developers is essential if the process is to go smoothly. Kasperson (1992) found that distrust toward developers' results in a long and inefficient process of negotiation and raises transaction costs.

Fairness

People's views regarding the fairness of the siting process has always been the

---

2 The hierarchical approach is the traditional scientific management approach. The decision maker first decides on the site using the technically based site screening and selection process and then announces and defends the decision. The voluntary-market approach involves such market-based instruments as compensation, economic incentives, and bargaining.

focus of the residents' appeals and it has been referred to in the literature to explain the NIMBY phenomenon for a long time (O'Hare et al., 1983; Linnerooth-Bayer, 2005).

Civic Duty

Civic duty is usually the reason for altruistic behaviours that can raise the well-being of fellow countrymen. Frey et al. (1996) and Frey and Oberholzer-Gee (1997) pointed out that civic duty explains why some residents support the establishment of NIMBY facilities even though they do not receive compensation.

Need

"Establish need" is one of the first few considerations in the list of the Facility Siting Credo authored by Kunreuther et al. (1993). Kasperson (2005) puts it in first place among such considerations as strategic imperatives for siting success. These authors argued that it is first necessary to demonstrate a clear need for the facility and to build a widespread recognition and consensus that the proposed siting is in the general public interest.

Social Pressure

Oberholzer-Gee and Kunreuther (2005) recognize that social pressure is particularly important for risky projects that impose sizable negative externalities on the residents of the host community. Under huge social pressure, the personal decisions of those residents are affected by the opinions of other residents in the community. They find that the extent of the personal support provided by one resident will be affected by other residents' opinions.

In addition to the social pressure from the community mentioned above, based on the experiences in Taiwan, we believe that there is another, more important, social pressure, namely, the social pressure from the general public and the media.

Socioeconomic Factors

In addition to the factors mentioned above, socioeconomic factors including income, educational level, sex, occupation, and other personal experiences are thought to be related to risk perception and risk acceptance in the literature. For example, Lesbirel (1998) finds that communities appear to accept a facility more readily when per-capita incomes are rising relatively rapidly. Thus, socioeconomic factors usually need to be controlled to keep other things equal in the analysis.

CASE STUDY

The household is the unit of our survey of Wu-chiu residents. Since most of the residents live in Taiwan and seldom travel to Wu-chiu, we not only visited residents in Wu-chiu in person but also mailed questionnaires to and telephoned the residents of Wu-chiu who lived in Taiwan in 2000. Of all the 93 households registered in Wu-chiu Township, we reached 65 households, and of these, 52 households allowed us to conduct interviews.

This paper focuses on the impact of compensation on the extent of the residents' acceptance of the facility. In this regard, we propose testing the following three hypotheses:

(1) Compensation raises the public's support for the building of the facility.
(2) If we do not limit the way in which the compensation is used, we can increase the public's support for the building of the facility.
(3) The respondent tends to support the facility if he thinks he is a representative of the residents.

To test these three hypotheses, we use a respondent's answers to related questions in the questionnaire to build three variables that represent the respondent's acceptance of the facility under three different situations, namely, (1) whether there is compensation, (2) whether there are restrictions on the use of the compensation, and (3) whether the respondent is one of the representatives of the Township. The dependent variable, the respondent's acceptability of the facility, is an ordered categorical variable. There are five levels of acceptability to be chosen by the respondent. We estimate the ordered categorical choice model of the resident's acceptability of the facility using the Gompit model under the PROBIT procedure in SAS.[3]

We adopt the following regression model to test the three hypotheses:

$$Oi=\beta_1 Dxi + \beta_2 Fi + \beta_3 Ti + \beta_4 Ni + \beta_5 SPi + \beta_6 Pi + \beta_7 Ri + \beta_8 Ii + \beta_9 Ei + \beta_{10} Li + \varepsilon i$$

[3] The Gompit model is based on the Gompertz distribution which is strongly negatively skewed. This model is chosen because it is close to the distribution of the dependent variable of our survey data.

where Oi is the resident's willingness to accept the facility with five ordered categories: Fi measures the respondent's valuation of fairness with five degrees; Ti measures trust in the developers with five degrees; Ni measures the need for the facility with five degrees; SPi measures social pressure based on his estimate of the share of nonsupporters; Pi is the respondent's view of the probability of the facility being built in the future, with five degrees of probability; Ri is the risk perception with five degrees; Ii is income; Ei is the educational level; and Li is a dummy variable that has to do with leaving Wu-chiu entirely. Dxi is the independent variable for the three hypotheses that is used separately in each model, namely, whether there is compensation, whether there is a limitation on how to use the compensation, and whether the resident being interviewed thinks he is a representative of the residents.[4]

Table 2 presents the estimates of the Gompit regression. The results of the first two models are as we expected. In the case of Wu-chiu, both the existence of compensation and using the compensation without restrictions can raise the acceptance of the facility significantly. The third model shows that being the representative of the Township has little bearing on whether the residents support the facility. This may be due to the fact that in this small and closely connected community the standpoint of Township representatives cannot be different from that of the residents even though they have experienced many more lobbying activities from both the developer and government agencies. The three regressions show that both the *fairness of the siting process* and *trust* in those developers have a significant effect on the acceptance of the facility as was expected in the literature.

However, what was not expected was finding that both the *need for the facility* and *social pressure* are insignificant explanatory variables. This may be due to the fact that most residents believe that they do not have to bear the civic duty to provide the site as a repository for nuclear waste as Wu-chiu has already borne a significant share of civic duty as a military base for the last sixty years. Thus, they think that they deserve the compensation and they do not care if the facility is needed by their fellow countrymen. Furthermore, the social pressure from the community faced by the residents is not an important factor because they almost reach a consensus in regard to both variables: the willingness to accept the facility and their estimate of the share of nonsupporters.

In the three models, risk perception is barely positively related to the acceptability of the facility. It is significant in the second model only. This may

---

[4] We measure the social pressure from the community following Oberholzer-Gee and Kunreuther (2005). However, the social pressure from the general public and the media is not used in the regression due to data unavailability.

be explained by the fact that the residents do not have to shoulder the risk of the facility, as most of them live in Taiwan now and the remaining residents would all move to Taiwan once the facility is built.

Table 2. Regression Models Explaining the Acceptability of the LLRW Repository

Independent variables	Acceptability of the facility		
	(1)	(2)	(3)
	Estimate (standard deviation)		
Constant	–8.99 (1.64)**	–3.73 (1.23)**	–3.66 (0.94)**
Compensation (1 = yes, 0 = no)	0.95 (0.35)**		
Offer compensation with limited usage of the compensation (1 = yes, 0 = no)		–2.12 (0.36)**	
Representative (1 = yes, 0 = no)			–0.15 (0.47)
Fairness ("1 = very low" to "5 = very high")	0.88 (0.23)**	0.33 (0.21)**	0.24 (0.16)**
Trust ("1 = very low" to "5 = very high")	0.86 (0.26)**	–0.12 (0.23)	0.31 (0.18)**
Need for the facility ("1 = very low" to "5 = very high")	0.08 (0.20)	–0.07 (0.16)	0.04 (0.14)
Social pressure (%)	0.004 (0.0086)	0.0003 (0.006)	-0.001 (0.005)
Risk perception ("1 = very low" to "5 = very high")	0.07 (0.17)	0.27 (0.16)**	0.02 (0.12)
Income (NT$10 thousand/year)	–0.003 (0.002)*	0.0001 (0.002)	–0.001 (0.001)
Level of education (7 levels)	–0.02 (0.11)	0.02 (0.08)	–0.01 (0.06)
Leaving Wu-chiu (1 = yes, 0 = no)	0.43 (0.41)	0.44 (0.29)**	0.12 (0.24)
The probability of the facility being built in the future ("1 = very low" to "5 = very high")	0.14 (0.17)	0.22 (0.16)*	0.05 (0.12)
N	78	81	160
Log likelihood	–77.75	–86.77	–218.01

Note: ** = significant at 95% level; * = significant at 90% level

Figure 2. The Conceptual Model of Compensation Acceptance in Taiwan and Switzerland

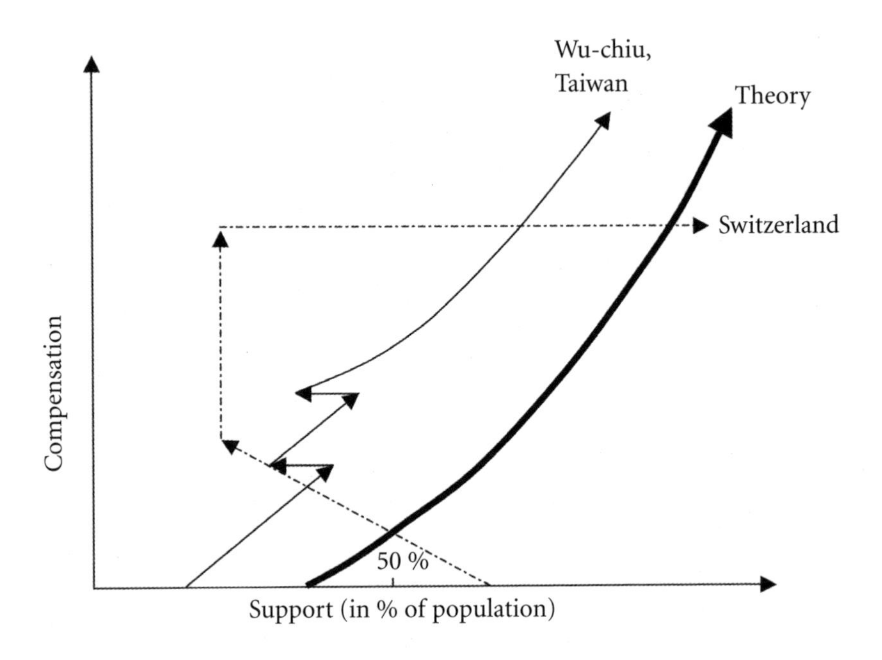

A NEW LOOK AT THE COMPENSATION CYCLE

Frey et al. (1996) ingeniously found a compensation cycle that depicts the relationship between compensation and the acceptance of building the facility when they studied an LLRW repository siting case in Switzerland. The compensation cycle in Figure 2 shows that the acceptance rate falls first from 50.8 percent to 24.6 percent as the level of compensation rises, and even though the compensation is raised to a certain level, the degree of the support of the residents still remains unchanged due to the rejection of bribes that would crowd out a sense of civic duty and public spirit. However, the higher-level compensation finally wins the support of the residents.

We try to enrich the compensation cycle model by adding the observed behaviour of Wu-chiu residents in Figure 2. The solid line for the Wu-chiu case in Figure 2 depicts the relationship between compensation and the willingness to accept the facility in Wu-chiu between 1995–2000. It looks like a path running zigzag up a hill. The genuine relationship is positive because of the crowding-in effect of compensation. However, we find that the residents' acceptance of the facility has been reduced several times mainly due to media exposure of the case. Thus, the social pressure from the general public and the media is much stronger than the social pressure from the community in Taiwan.

Although the positive relationship between compensation and the willingness to accept the facility is the same in these two cases, there is one major difference between them, i.e., the sense of civic duty before the siting process. Wu-chiu's residents do not think they have a civic duty to host the LLRW repository because the nuclear waste was not produced by them and they have served the country as a military base for the last sixty years. However, most residents in Switzerland regard it as their national duty to accept the repository. Thus, the ordinary level of the residents' civic duty may be the key factor to determining the compensation effects. When residents have a lower sense of civic duty, as in Wu-chiu, compensation has a crowding-in effect. However, if residents have a higher sense of civic duty, as in Switzerland, the compensation effect becomes a crowding-out effect.

Although we are not able to observe the final outcome of the Wu-chiu case because Taipower dropped its further development during the third phase of the voluntary siting process because of the change of the government in 2000, we expect that if the Wu-chiu case could proceed to the next phase, the higher level of compensation would finally win the support of the residents (see the dotted line in Figure 2).

POLICY IMPLICATIONS

This chapter finds that compensation would enhance the residents' acceptance of the facility in the long run. However, the compensation effect could be complicated by other factors such as civic duty, constraints on using compensation, fairness, trust, and social pressure in the short run. Among them, civic duty is particularly important. On the one hand, the Switzerland case shows that residents with a strong sense of civic duty would tend to host the facility for the benefit of fellow countrymen (Frey et al., 1996). However, Frey et al. (1996) also discovered that a strong sense of civic duty and compensation would counteract each other in the short run because of the crowding-out effect. On the other hand, the Wu-chiu case shows that, if the residents' sense of civic duty is weak, they would agree to host the facility only when enough compensation is offered. In the long run, compensation prevails.

Two policy implications can be drawn from these findings. First, the society as a whole and its leaders in particular should pay greater attention to building the society's social capital, of which trust, fairness, and the sense of civic duty are important parts. Social capital consists of those features of social organization, such as trust, norms, and networks that can improve the efficiency of society by facilitating coordinated actions (Putnam, 1993). A society with a strong social capital would make the siting process operate more smoothly. Second, even though compensation is effective, it should be offered only when necessary, such as when the residents' sense of civic duty to host the facility is weak.

For developers, it is important to formulate the siting process and the compensation package without the crowding-out effect. Developers should make the siting procedure fairer and more open to the public right from the beginning and let the public get involved in the siting process earlier. To increase trust between the public and developers, door-to-door visits may be better than only focusing on lobbying the leaders of the residents. The developers should not impose limitations on the use of the compensation.

REFERENCES

Baumol, W. J., & Oates, W. E. (1988). *The theory of environmental policy*. New York: Cambridge University Press.

Frey, B. S., & Oberholzer-Gee, F. (1997). The cost of price incentives: An empirical analysis of motivation crowding-out. *American Economic Review, 87*(4), 746–755.

Frey, B. S., Oberholzer-Gee, F., & Eichenberger, R. (1996). The old lady visits your backyard: A tale of morals and markets. *Journal of Political Economy, 104*(6), 193–209.

Jenkins-Smith, H. C., & Kunreuther, H. (2005). Mitigation and benefits measures as policy tools for siting potentially hazardous facilities: Determinants of effectiveness and appropriateness. In S. H. Lesbirel & D. Shaw (Eds.), *Managing conflict in facility siting*. Cheltenham, UK: Edward Elgar.

Kasperson, R. (2005). Siting hazardous facilities: Searching for effective institutions and processes. In H. Lesbirel & D. Shaw (Eds.), *Managing conflict in facility siting*. Cheltenham, UK: Edward Elgar.

Kasperson, R., Golding, D., & Tuler, S. (1992). Social distrust as a factor in siting hazardous facilities and communicating risk. *The Journal of Social Issues, 48*(4), 161–187.

Kunreuther, H., & Easterling, D. (1996). The role of compensation in siting hazardous facilities. In D. Shaw (Ed.), *Comparative analysis of siting experience in Asia*. Taibei: Academia Sinica.

Kunreuther, H., Fitzgerald, K., & Aarts, T. (1993). Siting noxious hazardous facilities: A test of the Facility Siting Credo. *Risk Analysis, 13*, 301–318.

Lesbirel, S. H. (1998). NIMBY politics in Japan: Energy siting and the management of environmental conflict. New York: Cornell University Press.

Linnerooth-Bayer, J. (2005). Fair strategies for siting hazardous waste facilities. In H. Lesbirel & D. Shaw (Eds.), *Managing conflict in facility siting*. Cheltenham, UK: Edward Elgar.

O'Hare, M., Sanderson, D., & Bacow, L. (1983). *Facility siting and public opposition*. New York: Van Nostrand-Reinhold.

Oberholzer-Gee, F., & Kunreuther, H. (2005). Social pressure in siting conflicts: A case study of siting a radioactive waste repository in Pennsylvania. In H. Lesbirel & D. Shaw (Eds.), *Managing conflict in facility siting*. Cheltenham, UK: Edward Elgar.

Shaw, D. (2005). Vision of the future for facility siting. In H. Lesbirel & D. Shaw (Eds.), *Managing conflict in facility siting*. Cheltenham, UK: Edward Elgar.

Shaw, D., & Shaw, R. (1991). The resistibility and shiftability of depletable externalities. *Journal of Environmental Economics and Management, 20*(3), 224–233.

Slovic, P. (1993). Perceived risk, trust, and democracy. *Risk Analysis, 13*(6), 675–682.

13

NIMBY: Environmental Civic Society and Social Fairness in China

Yang Yan

NIMBY IN CHINA: FROM NAUGHT TO EMERGENCE

NIMBY has received considerable attention in recent years. The term NIMBY (Not in My BackYard) refers to the exercise of environmental rights by residents in resisting the establishment of public facilities and services, which may damage the local environment in their neighborhoods (Mazmanian & Morell, 1992). Those residents typically employ formal and informal means to block the construction of these services. Public services that are susceptible to NIMBY include highways, homeless shelters, mental institutions, prisons, power plants, crematories, garbage dumps, nuclear reactors, and waste dumps. NIMBY facilities provide a full or partial social function. However, these facilities may affect the neighboring environment, threaten the residents' quality of life, undermine real-estate prices, and even drive away businesses; residents are typically worried about the damages brought about by such facilities on both the local environment and economy. Furthermore, there are concerns that NIMBY facilities may tarnish the image of local communities. These negative impacts on the communities' image may trigger feelings of inferiority among the local residents. It is therefore not difficult to understand the grievances associated with NIMBY and why the residents would take collective action to resist the construction of NIMBY projects (Tang & Weng, 1994).

The term "NIMBY" in China carries a special meaning in the process of development. The notion of NIMBY emerged in China during the process of "opening up," at a time when state control of society is undergoing transformation, and civil society, broadly defined, has begun to emerge. From the period of 1949–1980, NIMBY was virtually nonexistent in China; even if it had existed, such behavior would not have been identified. In other words, the conditions in prereform China did not favor the existence of NIMBY due to political control over society and state-dominated environmental governance. The Chinese cultural traditions also prevented NIMBY from developing. Among these factors, the key factor for the underdevelopment of NIMBY in China is the absence of the notion of environmental rights.

First, let us consider structural obstacles against the emergence of NIMBY in China: political control and state-dominated environmental governance. Before 1978, China was under a totalitarian regime. The ruling party exerted tight control over society through dense grassroots organizations, leaving almost no space for social activities and initiatives. Environmental degradation was regarded as a characteristic outcome of capitalism. Exploitation of the environment was not supposed to happen in a socialist country. The notion of environmental rights was thus fundamentally rejected.

Following reforms and liberalization in the post-1978 period, environmental work began to gain attention. On December 31, 1983, environmental protection was established as a keystone policy for state development in the Second National Environmental Protection Convention held in Beijing. This heralded the arrival of a new stage of environmental management. Subsequently, the state began to acknowledge the rights of citizens over the environment (but it was not until 1983 that a legal system for environmental regulation was established).

At present, the system of environmental governance in China is quintessentially state-dominated. The state alone sets policies, mobilizes resources, and implements rules. The role of the state and the society is severely imbalanced in environmental protection. Such imbalances in turn sustained the state's domination over environmental governance. Social exclusion from the environment is manifested in many ways (Xia, 2008). First, the government monopolizes information about the environment. Transparency of information is crucial for citizens to enforce their environmental rights; civic concern in turns pressures governments and businesses to fulfill their environmental responsibilities. Other than through governmental reports, the Chinese public has extremely limited access to information on policy making, construction projects, and industrial pollution. The public is forbidden access to important pieces of information on the environment. For example, one has to seek approval from the environmental monitoring authority for the results of any examination of pollution victims conducted by the environmental monitoring department. Second, societal influence on environmental policy making is currently restricted to experts and scholars. The final decision making remains in the hands of the government. The public is still unable to express its opinions or influence policy making directly. Third, with respect to environmental disputes and compensation, the environmental legislation system in China picked up the pace only in the reform era. However, most of these rules and regulations are largely made by environmental authorities; few go through formal legal channels. In mediating cases where actions afflict the public good, for example, China adopts an administrative remedial process, which relies mostly on public authorities to resolve disputes. This presents an obvious problem because in many cases the public authorities are the very participants of the actions in dispute. In these instances, putting public environmental rights aside is the natural choice for the government. Unorganized victims face tremendous difficulties safeguarding their rights. Needless to say, in a context of environmental governance dominated by the state, few opportunities exist for NIMBY to emerge.

Another reason against the development of NIMBY in China is the Chinese cultural traditions. The absence of NIMBY in China and the weaknesses of civic environmental rights can be traced back to China's long history and culture. In traditional China—and even until these days—the ways to handle business are highly stratified and hierarchical, showing great deference to rulers. This allowed the state's claim of "public interests" to take on legitimacy. Nonstate organizations cannot claim to represent the true public interests. Under such cultural influences, the public has a natural dependence on the state. Socialism and central planning further institutionalized individuals' reliance on the state.

Under the socialist system, the state monopolized all resources and took charge of the provision of all social services. To resolve any collective or social problem, the people had to turn to the government. Hence, the public's concern about and advocacy on environmental issues is to a larger extent dependent on mobilization by and the response of governmental authorities. Whenever an environmental problem arises, the first solution that comes to the public's mind is the government. The Chinese people typically place their hopes on the government for solving problems. The more this is the case, the more the people are unaccustomed to using other avenues of problem solving, and the more reliant they are on the state. Such reliance is articulated in myriad ways. First, the public tends to trust information provided by the bureaucracies and quasi-bureaucracies rather than by nonstate organizations. Second, the public believes that the success of environmental protection relies upon the perfection of laws, education by the state, and governmental investments. When faced with environmental problems, the public lacks initiative in participation and problem solving. In short, under such cultural and institutional conditions, any private or organizational efforts to enforce one's environmental rights will be constrained by public opinion and political control, nipping the emergence of NIMBY in China.

NIMBY IN CHINA: ENVIRONMENTAL RIGHTS AND SOCIAL JUSTICE

As reforms and liberalization proceed, the state is gradually loosening its control over society. Together with worsening industrial pollution, environmental protection has become a key concern of government. Simultaneously, the Chinese public is gradually transitioning away from the lack of civic consciousness and reliance on the state for solutions under socialist rule. Hence, we observe changes happening in environmental governance both in the structure of government and in civic consciousness. It is under these conditions that NIMBY begins to appear. Since its inception, NIMBY in China has transcended the realm of environmental activism arising from the selfish interests of local residents. Chinese activists had furthermore turned environmental crises from technical into policy-making issues. Their objective is to defend social justice, which is the defining feature of NIMBY in China. We can analyze this defining feature using the following case studies.

The dam construction incident in Nujiang, Yunnan Province, and the PX (p-xylene) incident in Xiamen are two of the most controversial and landmark events in China. They are representative of NIMBY in China. By examining their points of controversy, major actors, and values, we can analyze the defining features of NIMBY in China.

Hydro-Electric Power Development in Nujiang—A Bottom-Up NIMBY Movement

In August 2003, considerable attention among environmental experts was raised when the plans for developing hydroelectric power in Nujiang River basin were promulgated. Two opposing camps emerged, with ecological, forestry, and geological experts on one side, and bureaucrats from the hydroelectric bureaus, planning commissions, and local government on the other. Environmental experts believed that developing hydroelectric power in Nujiang would affect ecological diversity, leading to soil erosion and damaging cultural relics.

Proponents of the development project, on the other hand, advocated that developing hydro energy would solve people's "livelihood" problem. Second, from the perspective of energy production, the Nujiang project could produce up to 22 million kilowatts. In comparison, the Three Gorges Dam produces 18 million kilowatts. Nujiang's investment, however, would cost only half of that of the Three Gorges Dam because of Nujiang's favorable geographical position, which makes it possible to produce more energy with less investment. Third, Yunnan is an ethnic minority region, in which 92 percent of the population is composed of ethnic minorities. The government authorities favored bringing economic development into this minority area. Fourth, policy makers had to consider maintaining friendly relations with the South East Asian countries to assure an energy supply that was under threat. There was an urgent need to develop new energy sources. If the dam were constructed, the hydro energy could be exchanged for oil. Part of the oil could be shipped to eastern and central China, and the other part to Myanmar, Pakistan, and India. Yunnan would serve as a conduit. From an international relations perspective, Yunnan was located in a strategic position.

The Yunnan local government and the National Development and Reform Commission (NDRC), formerly State Planning Commission (SPC) and State Development Planning Commission (SDPC) approved the Nujiang dam construction project. On June 14, 2003, Yunnan Huadian Power Corporation Limited, a hydroelectric company, was established. On July 18, the Yunnan Nujiang Dam station was officially established. On August 14, NDRC approved the "Nujiang's lower basin hydro-electric power developmental report," which once again stirred controversy. Dam construction in Nujiang faced opposition from the State Environmental Protection Administration (SEPA), residents of the Nujiang lower basin, and various environmental experts primarily experts in Beijing, for example, including the founder of the Friends of Nature, Liang Congjie 梁從誡, who is also a member of the national committee of the Chinese People's Political Consultative Conference (CPPCC), as well as Professor He Daming, the leader of the Yunnan river project and director of Asia International Rivers Centre of Yunnan University.

There were two points of controversy concerning the Nujiang dam project. On September 1, 2003, the Law on Environmental Impact Assessment was officially promulgated. The controversy surrounding the Nujiang incident turned from a debate on the dilemma between environmental protection and economic development into the scientific basis and procedural justice of Nujiang environmental impact assessment report. In September 2003, SEPA requested to review the Nujiang environmental impact assessment report. A month later, the Yunnan authorities offered a compromise policy with regard to the project's environmental impact and planning. In light of the two experts' roundtables organized by the SEPA, the Yunnan Environmental Bureau conducted two of its own conferences on September 29 and October 10 to reassess the feasibility of the Nujiang project. On January 5, 2004, the Nujiang project's environmental impact assessment report prepared by the State Hydroelectric Investigation Design and Research Institute was approved by SEPA. On February 18, 2004, Premier Wen Jiabao commented, with reference to the "Nujiang lower basin hydro-electric power development and planning report," that "on such major projects that are highly regarded by society, especially with opposing views on their environment impact, we should conduct research cautiously and make scientific decisions."

The Nujiang dam project was temporarily postponed. The environmentalists appeared to have won a victory. After the Nujiang project was ordered to cease, the environmentalists directed the controversy to the content and scientific basis of the Nujiang environmental impact assessment report. Since these documents were not made public, citizens did not know how the developers and local government avoided issues relating to environmental damages, relocation of residents, safety, and economics of the dam. After months of silence, scattered media reports revealed that the second conference of the experts was held on January 13, 2004 in Beijing and, according to a Chinese NGO (*Qingxi nujiang*, 2005, August 25), many of them had only been informed shortly before the meeting and many had to rush to Beijing for the meeting.

It was only at the meeting that environmental documents and reports prepared by the National Power Company Investigation and Design Institute and East China Investigation and Design Institute were distributed but they were promptly taken away at the end of the meeting. Accordingly, the attendees were unable to even recall the full name of the conference. Environmentalists protested this incident, claiming that such a process of decision making did not meet the demand when major policies ought to have public participation and legal backing. The way the matter was handled neither met international norms nor complied with the Administrative Permission Law nor the State Council's "Provision on implementing policies according to law" which advocated the transparency and availability of information to the public.

According to the Environmental Impact Assessment Law that was promulgated on September 1, 2003, "the state encourages all related departments, experts, and members of the public to participate in the environmental impact assessment in appropriate ways." It also stipulates that "specialized departments that are planning projects that could have potential effects on the public's environmental rights should conduct participatory meetings and conferences to seek opinion of related departments, experts, and members of the public before such plans are submitted for official approval." At the same time, "[these departments] should seriously consider the opinions of related departments, experts, and members of the public, and when submitting the plans for approval, the departments should append an explanation of why they have or have not adopted the suggestions put forth."

On August 10, 2008, a directive entitled "Temporary administrative provisions on environmental protection hearing" was issued by the Ministry of Environmental Protection, listing two types of construction plans and over ten kinds of special projects, about which the participatory decision-making process should be implemented. The two major construction projects include: (a) projects that may bring substantial environmental damage to medium-to-large sized projects that require the issuance of environmental impact assessment reports; and (b) small-scale projects that may produce smoke, odors, and noise, which may affect the living environment of the neighboring residents. The ten major special projects refer to those that may cause environmental damage or that may impinge on the environmental rights of residents, including projects in manufacturing, agriculture, livestock, industry, forestry, energy, water resources, transportation, urban development, tourism, and natural resource development. Obviously, the Nujiang dam project is one of those projects that requires public hearings.

The Nujiang dam incident reveals many typical characteristics of NIMBY projects, but it also has its uniqueness. First, the initiators of the movements are environmental groups including the environmental nongovernmental organizations (ENGOs), rather than local residents. The movement was first initiated by the founder of Green Earth Volunteers, Wang Yongchen. On learning about the project from officials at the State Environmental Protection Administration, Wang contacted Professor He Daming of Yunnan University and they jointly got under way an environmental movement. ENGOs have become an important force in uniting public environmental consciousness in an organized fashion in influencing policy making in a rational and orderly manner. ENGOs also serve to fill the gap in environmental protection brought about by "governmental failure" and "market failure."

From an organizational perspective, the Nujiang incident appears to be a social environmental movement. It indicates that Chinese society is gradually transforming from a state-directed society to one in which citizens can exert their rights. In the face of environmental crises, citizens begin to organize themselves and participate actively in environmental groups. We also observe that the controversy behind the Nujiang Incident shifted from a debate on the "environmental damage versus economic value" to the scientific basis and procedural justice of the environmental impact assessment process. As a NIMBY phenomenon that has emerged in China's transitional society, the points of controversy illustrate that NIMBY in China is not only opposed to the policy makers' decisions, but also to the issues and problems framed by the experts engaged by the government. These actions come not from self-interest, but from the determination and will to protect one's homeland. More importantly, it represents a deep mistrust of government, particularly on the government's approach to resolving such problems. Such mistrust comes, in turn, from the awakening of environmental consciousness in China.

The Xiamen PX Incident: A NIMBY Movement Initiated by Intellectuals and Local Residents

The case in 2007, which caught national attention, is another landmark case of environmental movement in China. The Taiwanese Xianglu Group (翔鷺集團) planned to construct a p-Xylene (PX) factory that would produce 80 million tons of PX each year. It was estimated that such a plant could contribute 80 billion RMB to Xiamen city's GDP annually. In February 2004, the State Council approved the project. In July of the following year, the State Environmental Protection Administration endorsed the Environmental Impact Assessment (EIA) Report of the project. In July 2006, the NDRC approved the project's application, and in November, the PX project commenced without delay. Yet, the corresponding Regional Environmental Impact Assessment (REIA) had not even begun. Many local residents were unaware of the project. In March 2007, at a session of the Chinese People's Political Consultative Council, Professor Zhao Yufen (趙玉芬院士) of the Chinese Academy of Sciences and Xiamen University, supported by 105 other members of the Council, submitted a bill asking for relocation of the PX project. The bill cast doubt about choice of location and safety of the plant. This became the most important bill of the year. The media pursued the story and the PX incident soon caught a lot of attention from the public.

In May, cell phone messages about "the stroll" spread quickly across Xiamen. Heated debates took place on the Internet. Various dedicated web sites, chat boards, and QQ (an instant-messaging software in China) groups proliferated. On the Internet, the Xiamen incident received the greatest number of hits during that time. The Xiamen Environmental Protection Bureau and the investors defended the PX project on the Xiamen Evening News and on the web site of the Xianglu Group. On May 30, the Xiamen government announced the postponement of the PX project. On June 1 and 2, Xiamen residents conducted a rational peaceful "strolling" (demonstration) with yellow bands on their wrists. The articulation of public opinion reached its peak. On June 12, the Environmental Protection Bureau organized a team of experts to conduct a new environmental impact assessment of the project, as well as the surrounding region. On December 5, the Xiamen government announced findings of the environmental impact assessment report and opened the issue for public discussion for ten days. During the period of December 13–24, 2007, the Xiamen municipal government selected 100 residents to participate in the experts' forum on the environmental impact report. Residents were allowed to have a direct dialogue with the government officials. Following the forum, the Fujian Provincial Government and Xiamen municipal government decided to relocate the project to the Gulei 古雷 Peninsula on Zhangzhou city. The environmental movement finally won a victory, reinforcing the role of public participation in the decision-making process of a major construction project (Zhu & Su, 2007, December 20).

Led by Professor Zhao Yufen, the PX campaign gained a lot of public support, including 105 members of the National Committee of the Chinese People's Political Consultative Council (CPPCC), most of whom were intellectuals. When the project was initially approved, local residents were not aware that the project had not gone through the Environmental Impact Assessment process and were not informed of the environmentally harmful effects of PX. Only those intellectuals with close contact with the major decision makers had heard about the project. Spearheaded by intellectuals, the local environmental movement, propelled by local groups such as the Xiamen Greencross Association, gained much momentum. Public sentiments focused on the government's decision-making process and many asked for the right to participate in the process. The participation of residents' representatives in the final deliberation, as earlier described, advocated for upholding the principles of democracy and social justice in public decision making.

From the two case studies presented above, NIMBY in China has the following characteristics:

(1) Although NIMBY was triggered by environmentally harmful construction projects, the movements opposing these projects went beyond parochial interests. NIMBY in China does not necessarily connote selfishness. It describes local residents' resistance to the construction of essential public services that was imposed by the government on certain localities. However, it is not hard to observe from the two cases that NIMBY in China has gone beyond protecting local interests. This feature has been manifested in the nature of the controversy surrounding the two incidents. Public dialogues about both the Yunnan dam project and the Xiamen incident show an ideology that eschews traditional pursuits for "development," adding an ecological depth to the grassroots movement.

(2) The NIMBY movements were directed at the procedural justice and legitimacy of decision making. Conventional NIMBY movements often involved the problems of ecological and monetary compensation. However, monetary gains or losses were never raised in the Yunnan and Xiamen

incidents. Instead, attention was placed on the potential environmental harms, as well as the procedural appropriateness of the environmental impact assessment and policy-making processes. In other words, the core of NIMBY in China is environmental value, social justice, and democracy.

According to the principles of social justice, all governments should ensure that all their citizens have the right to fair treatment. In other words, in a just society, every person should have equal access to resources to pursue his or her own interests or goals. Christiano (1997, p. 67) and Dworkin (1978, p. 272) called on governments to treat each person with "equal care and respect" and to put citizens' welfare first. On the one hand, it was suggested that the government should leave each person to decide his or her own interests and conception of the good life. However, as each person is highly dependent on others and as we have divergent views on how best to distribute resources, the only way to ensure that interests are met is through institutions with coercive power. Hence, to ensure that each person's rights are not harmed, it is necessary to grant all citizens an equal right to opportunities and access to influence collective decision making. Principles of justice are founded upon considerations of equality. Equality, in turn, depends on the mere fact that each person should have equal resources to comprehend and pursue his or her interests.

Where a conflict of interest exists, every citizen should be given equal resources to participate in the decision-making process. This implies the principle of political equality, including but not limited to the right to vote in decision making. Moreover, political equality also requires that each citizen has the opportunity to voice his or her opinion and influence the collective decision-making process. Such a process is necessarily a democratic one.

Especially in dealing with the problem of resource allocation, participatory decision making should be conducted. The will of the people should be articulated in the political process itself. China's NIMBY movements have, from their inception, pursued the goals of procedural justice and political fairness. Xue Ye, the environmentalist and head of Friends of Nature, once said, "Whether or not a dam should be constructed in Nujiang is at the early stage of research and discussion. The first important thing is transparency of information to be allowed by the government." More precisely, he stressed, "First, we demand openness, so that the public is aware of the situation; second, fairness implies justice, equality between strong and weak groups, between the western and eastern regions, between city and countryside, and between humanity and nature should be taken into account. Third, we demand adherence to China's laws and regulations. Fourth, we insist on sustainable development. If we follow these four principles, we would be able to accept the outcomes, no matter how they turn out. If we follow these four principles, we believe the ultimate policy will be more rational, perfect, and acceptable to society. If we work against these principles, then the developmental policies will provide opportunities for rent-seeking, corruption, and haste to capture interests" (China Energy Net, 2005, July 12).

(3) NIMBY in China aims to establish a conflict-resolution mechanism through democratic participation, on the promise of environmental and social justice. Dams and similar projects are public goods that may generate benefits for those who enjoy these services and incur costs to some, particularly local communities, who bear them involuntarily. Hence, a society needs to develop a restraining mechanism for collective decision making on these types of goods. As it is not easy to reach a consensus, conflicts are inevitable. Hence, the question of developing a fair redistributive outcome becomes the most important issue in resolving the problem.

BEHIND NIMBY: CHINA'S ENGOs IN ACTION

China's NIMBY movements reflect the relations between the state and society in a broad sense, the most important of which is the rise of the environmental civil society. The environmental movement in China is composed of participation by citizens with mobilization by environmental nongovernmental organisations (ENGOs). The most direct form of participation is through collective action, as seen in the Yunnan and Xiamen incidents. The ultimate goal of China's environmental movement is to bring about better governance. In other words, NIMBY in China poses a challenge to the existing regulatory environment and political decision-making process, in order to bring about environmental fairness.

In the traditional environmental management system, citizens could participate in environmental decision making through environmental petitioning, making suggestions to the National People's Congress and National People's Consultative Councils, and participating in dialogue sessions initiated by the State Environmental Protection Administration (SEPA). According to official statistics from SEPA, the number of petitions grew dramatically in the 1990s. In 2004, the number of letters hit a peak of 595,852, while the number of visits reached 86,414. Environmental bills put forth by the National People's Congress and National People's Consultative Councils reached 12,532 (Ministry of Environmental Protection of the People's Republic of China, 2005). Some new forms of environmental participation have also emerged in China, including discussion sessions in the environmental impact assessments of major projects, Internet forums, administrative litigation of inappropriate agency actions, civil law suits against polluting industries, and intervention into state decision making by environmental experts (Shi & Zhang, 2006).

The Yunnan and Xiamen incidents took place against a larger backdrop of increased social participation in environmental petitioning and litigation. Other similar incidents include the "stroll," a form of spontaneous "march" without formal organization, in Shanghai and in Pengzhou of Chengdu city. These participants got together through the Internet and cell phone messages, forming collective environmental movements. These are signs of the rise of environmental civil society and new forms of social participation in environmental decision making.

Another important element of the environmental civil society is the formation of ENGOs, which make environmental protection their missions, are nonprofit and nonadministrative organizations, and provide an environmental service to the public. (All-China Environment Federation, 2006) Since the establishment of the Chinese Society for Environmental Sciences, China has witnessed the formation of 2,768 ENGOs, of which 1,382 (or half) were created by the government. Those initiated by citizens numbered 202, or 7.2 percent of the total while those created by students and other collective bodies made up 1,116, or 40.3 percent.

The influence of non-state-initiated environmental civic groups has been gradually increasing. They play a unique function in the environmental movement. In most cases, ENGOs have close relationships to the government, which gives them an edge in influencing policy making. In most cases, environmental government-organised nongovernmental organizations (EGONGOs) have a more direct impact on environmental decision making. In contrast to the social ENGOs, EGONGOs are absorbed into the governmental structure and most of their leaders, some retired senior officials who still receive government subsidies, are appointed by the state. Such an arrangement allows EGONGOs to influence policy making, while ensuring structural stability for the state. The unique relationship allows EGONGOs to directly participate in environmental affairs, in the formulation of rules and regulation and to gain access to information. In return, members of the EGONGOs are expected to render their expertise and advise the state on environmental matters (Schwartz, 2004).

China's NGOs are currently exerting greater and greater influence through collaboration. Many ENGOs contact each other through the Internet and forums, and are actively engaged in the discussions about "hot" topics (Yang, 2005). Among them, the Friends of Nature, despite not being an EGONGO, merits special mention. Its founder, Liang Congjie, is a public figure well known, for his own work as well as for being the grandson of the respected scholar Liang Qichao 梁啟超, a renowned thinker, political activist,

Figure 1. Number of Environmental Petitions, 1996–2006 (Source: State Environmental Protection Administration: www.zhb.gov.cn/plan)

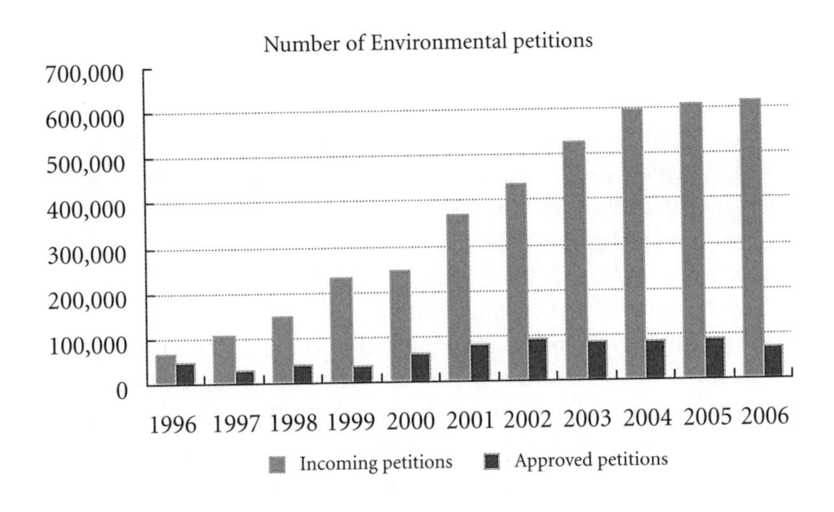

Figure 2. Composition of ENGOs in China, 1996–2006 (Source: All-China Environment Federation: www.acef.com.cn)

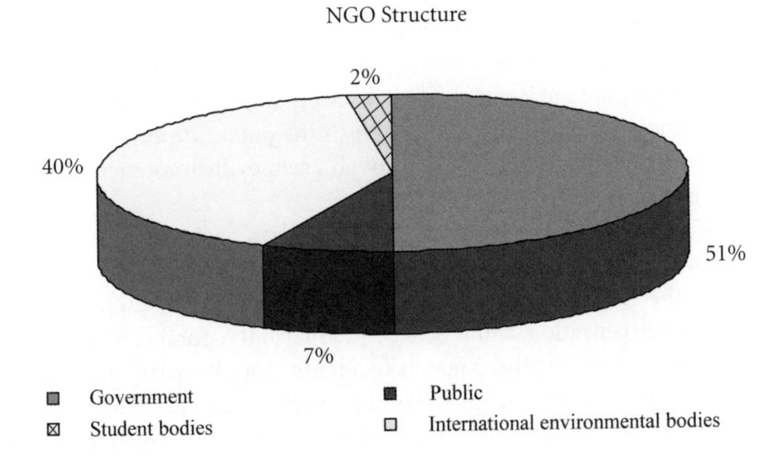

and litterateur in modern Chinese history. Highly regarded as a great master in Chinese academic circles, Liang Qichao has commanded much respect, just like his grandson. With such a high social status, the Friends of Nature was given a boost of enthusiasm and rigor when it collaborated with the Green Earth Volunteers, which comprised a group of scientists, state officials, and reporters. Hence, they have more avenues to access privileged information and participate in policy making. In fact, news about the Nujiang project was first obtained by senior members of Green Earth Volunteers from the State Environmental Protection Administration. In the course of the Nujiang incident, other environmental groups emerged, including the "Green River" (綠色江河, www.green-river.cn/Article/ShowClass.asp?ClassID=18), Green Home (綠色家園, www.greenhome.net.cn/english.htm), and the Green Watershed 綠色流域, www.greenwatershed.org/). Their joint efforts were so concerted and effective that their views swayed the Yunnan Provincial People's Consultative Council and the State Environmental Protection Administration. The latter subsequently invited 36 experts in the fields of ecology, agriculture, forestry, geology, water resources, fishery, wildlife protection, and cultural heritage protection to investigate the problem and formed a normative consensus to protect Nujiang and reconsider development.

As we can see from the discussion and as civil society emerges in China's transitional economy, China's environmental management has transformed from a state-dominated system to one with increasing public participation. China's environmental civil society, including environmental civil participation and ENGOs, has made impressive strides in recent years. As the government gradually makes way for greater public participation in environmental decision making, the environmental civil society has an increasing influence on China's environmental affairs. Members of the public are at the same time learning how to express their views through avenues such as environmental petitioning (Barry, 1996).

It can be expected that as more ENGOs appear, civil society is going to have a growing influence on environmental affairs. It is important to understand that the rise of civil society in China takes place in the context of reforms and marketization, and is hence a gradual and natural product.

In both the Yunnan and Xiamen incidents, popular participation and ENGO participation are the main forms of participation. In particular, ENGOs have played an indispensable role in mobilizing resources and creating a public space that made NIMBY possible. It is in the context of environmental civil society that the NIMBY movements are able to transcend from local self-seeking interests to take on the missions of promoting social fairness and procedural legitimacy for the great majority. These missions help to equalize environmental resource allocation in China, and to promote environmental egalitarianism and regulation. Another aim of these movements is to ensure that policy making takes place under specified rules. In short, improved governance implies improved interactions among state, market, and society. Improved governance rests on the principles of the rule of law, legitimacy, fairness, transparency, accountability, and participation. We can see that social participation is an important force in promoting improved governance. Participation by citizens and social groups are an important requisite for promoting sustainable development.

REFERENCES

All-China Environment Federation. (2006). *Blue book of development of Chinese environmental non-governmental organizations.* Beijing: China Environmental Science Press.

Barry, J. (1996). Sustainability, political judgement and citizenship: Connecting green politics and democracy. In B. Doherty & M. de Geus (Eds.), *Democracy and green political thought.* London and New York: Routledge Press.

China Energy Net. (2005, July 12). Dispute over the Nujiang River—pain in selection in development mode. Retrieved from http://www.china5e.com/news/water/200507120073.html

Christiano, T. (1997). The significance of public deliberation in Bohman. In J. Bohman & W. Rehg (Eds.), *Deliberative democracy.* Cambridge, MA: MIT Press.

Dworkin, R. (1978). *Taking rights seriously.* Cambridge, MA: Harvard University Press.

Hunold, C., & Young, I. M. (1998). Justice, democracy and hazardous siting. *Political Studies, 46,* 82–95.

Mazmanian, D. A., & Morell, D. (1992). *Beyond superfailure: America's toxics policy for the 1990s.* Boulder, CO: Westview Press.

Ministry of Environmental Protection of the People's Republic of China. (2005). National Environment Statistical Bulletin. Retrieved from http://www.zhb.gov.cn/plan/hjtj

Qingxi nujiang. (2005, August 25). Nujiang hydropower environmental assessment should be public in accordance with the law (in Chinese). Retrieved from www.nujiang.ngo.cn/Dynamics/2005/080

Schwartz, J. (2004). Environmental NGOs in China: Roles and limits. *Pacific Affairs, 77*(1), 28–49.

Shi, H., & Zhang, L. (2006). China's environmental governance of rapid industrialization. *Environmental Politics, 15*(2), 271–292.

Tang, C. P., & Weng, W.-T. (1994). Destructing Not-In-My-Back-Yard movement: Political mobilization in local protests against highway construction projects in Taiwan. *Taiwan Chengchi University—Journal of Public Administration, 14,* 125–149.

Xia, Z.-H. (2008). Shortage of environmental rights of China's public from the view of social exclusion. *China Population Resources and Environment, 2,* 49–54.

Yang, G. (2005). Environmental NGOs and institutional dynamics in China. *The China Quarterly, 181,* 46–66.

Zhu, H. J., and Y. T. Su (2007, December 20). Public opinion and wisdom change the fate of Xiamen (in Chinese). *The Southern Weekend.* Retrieved from http://blog.sina.com.cn/s/blog_4e21402f01007sjf.html

14

Conclusion

S. Hayden Lesbirel

This volume has explored the siting of unwanted facilities in the Asia-Pacific region. It has done this by analysing several cases in the context of a theory of knowledge production and utilisation and the relationship between the two. It suggests that siting has become a national policy issue in the region, despite the diversity of nations, and that there is a need to bring the Asia-Pacific siting experience more explicitly into the growing comparative literature on siting. The chapters of this volume reveal that the Asia-Pacific experience can contribute to a better theoretical and comparative understanding of siting and that it can provide useful insights for policy makers and practitioners on siting processes and their outcomes. They also illustrate that the literature makes several assumptions with respect to the relationship between spillover effects and the scope of the community and decision processes and outcomes. This chapter concludes that the literature needs to redefine the relationship between spillover effects and the nature of community and the relationship between those effects and decisions if we are to enhance our understanding and management of siting disputes.

KNOWLEDGE PRODUCTION

Siting has become a major social and political issue in the Asia-Pacific. Rapid economic growth in the region has increased the demand for a host of facilities such as waste repositories, energy facilities, a range of large-scale industrial projects, and even large entertainment projects such as casinos. At the same time, there has been significant democratisation in the region with political reforms that have ushered in more public participation in political and electoral processes. These two trends have elevated siting onto political agendas in nations in the region. Increased participation in political processes has led to significant siting conflicts in the region and this has the capacity to stall and even kill major projects necessary for continued economic growth and development.

Comparatively, siting has become a critical policy issue independent of the level of economic development and the form of government. As the chapters of this volume demonstrate, siting conflicts have occurred in more developed areas such as North America, Japan, South Korea, Singapore, Hong Kong Special Administrative Region (SAR), and Taiwan, as well as in developing nations such as China and Vietnam. Similarly, siting controversies have been witnessed in recognized democratic regions such as some of North America, Japan, South Korea, Hong Kong (SAR), and Taiwan, as well as in nations that are often viewed (as least by some) as less democratic such as China, Vietnam, and Singapore. The siting issue is omnipresent in the region independent of general political and economic classifiers in comparative social science.

As this volume attests, there has been a growth in the literature on siting issues in the region, reflecting the growing importance attached to this field of enquiry both as an analytic and as a policy issue. Although this literature is still dwarfed by the literature in the West, including that of Europe, the chapters, which were written by leading scholars in the region and which provide a reasonable indicator of the state of the literature in the region, suggest important theoretical and methodological characteristics.

As with the siting literature more generally, the siting literature in the Asia-Pacific is dominated by foundationalist approaches to ontology in the social sciences. These approaches posit that there is a real siting world out there and that it is independent of our knowledge of it. Epistemologically, and similar to the literature more generally, the Asia-Pacific literature is dominated by positivists who argue that it is possible to understand siting processes and outcomes through theory and by testing those theories by direct observation. Anti-foundationalist ontological approaches with interpretivist and realist epistemological positions have yet to make virtually any inroads into the siting literature in the Asia Pacific, compared to that in the literature more generally where we observe that those approaches are starting to gain some currency.

Within this context, the regional literature is characterized by a rich diversity of theoretical and methodological approaches. Mitchell along with Nguyen and Maclaren provide interesting institutional analyses of siting issues. Baxter, Lam et al., and Ishizaka et al. offer behavioural perspectives on siting conflicts. Kunreuther and Quah and Toh investigate siting from the perspective of rational choice theory. Others adopt more multitheoretic approaches to understanding siting processes and their outcomes. Chiou combines behavioural and institutional approaches, Shaw combines behavioural and rational choice approaches, while Aldrich and Yang integrate institutional and behavioural perspectives. There is a similar diversity in terms of methodological approaches. Mitchell, Baxter, Chiou, and Yang use mainly qualitative approaches, Shaw and Quah and Toh use mainly quantitative methods, while Aldrich, Kunreuther, Lam et al., and Nguyen and Maclaren employ more mixed methodological approaches. The siting literature in the Asia-Pacific, like that elsewhere, is clearly characterised by theoretical and methodological diversity and contestability within the dominant approaches.

Social scientists working on siting issues in the region have identified a range of key variables that assist in explaining recurring patterns of siting processes and their outcomes. They have identified such variables as risk, distribution of burdens, policy instruments (including compensation), social capital, trust, and legitimacy and their interactions as important factors influencing siting in the region. Similarly, they have developed a variety of middle-range theories that tend to focus on specific country contexts that are narrower in their explanatory power and more conditional in terms of their conclusions than general theories. The chapters contained in this volume cumulatively suggest that there is a broader set of variables likely to be crucial for explaining siting processes and their outcomes independent of culture. For instance, the levels of trust and social capital are just as critical in understanding siting conflicts in China, Vietnam, Hong Kong (SAR), and Japan as they are in the United States and Canada. Bringing siting experience from the Asia-Pacific explicitly into the broader literature on siting in the West and elsewhere will enhance our theoretical and comparative understanding of this increasingly important issue.

KNOWLEDGE UTILISATION

As noted by Lesbirel in this volume, siting scholars have generally attempted to provide policy-relevant analyses. The chapters in this volume individually and collectively reinforce that observation and highlight important insights for policy makers in the region. We examine briefly conceptual and instrumental (diagnostic, predictive, prescriptive, and evaluative) insights under three broad categories: distributive equity (procedural and substantive), public involvement (trust, sincerity, credibility), and policy instruments (compensation and mitigation).

Conceptually, an important conclusion is that coercive approaches to siting are becoming less effective in the region. Many nations in the past, including the United States and Taiwan, adopted more coercive approaches to siting using, for example, eminent domain. Yet, over time, these and other areas such as Canada, Japan, Hong Kong SAR, and Singapore have generally introduced more voluntary siting processes. Even within Communist states, such as Vietnam, as Nguyen and Maclaren note, community development regulation has increased community power. In response, the state has introduced EIAs and other measures for community participation, although some would argue those measures have not gone far enough. Yang notes similar developments in China, especially in managing the emergence of environmental NGOs. This suggests that there is some convergence with European approaches to siting and that policy makers in the region could usefully learn from that experience in the further refinement of approaches to siting unwanted projects.

Diagnostically, a key siting issue in the region, with the exception of Japan and North America (as more established democracies), relates to the management of conflict under changing social, political, and economic circumstances. Increased demand for projects as a consequence of social and economic growth in the region coupled with expanded demands for environmental quality and democratisation have, as they did in Japan, led to the emergence of siting conflicts in all nations in the region. Yet, governments, where they are undergoing democratic reform, have confronted significant challenges in dealing with these conflicts. In the past, they may have been less concerned with distribution justice, seeking to locate projects coercively on concentrated communities, without taking into account adequately community interests. While they may have encouraged some public participation, they have tended to be opaque in decision making and have made decisions behind closed doors. This may have led to a loss of credibility and trust in private and public organisations involved in siting. There may have been a lack of an institutional apparatus for dealing with citizens' interests and, together with a lack of access to key decision making circles, this may have intensified conflicts. Furthermore, there may not have been adequate policy instruments such as compensation and mitigation for dealing with the spillover effects from unwanted projects. Siting in the region is thus one critical indicator of the nature and extent of governance.

The chapters in this volume offer useful predictive insights about siting. Mitchell predicts that environmental equity and justice issues are likely to emerge even with voluntary siting as projects are likely to be located in poorer communities. Baxter argues that intercommunity conflicts will emerge where developers do not address broader regional equity and fairness issues. Chiou posits that trust and credibility issues are likely to exacerbate confrontation at election times as local politicians become more heavily involved in siting. Lam et al. suggest that risk perceptions are likely to be amplified and complicate siting where effective public participation strategies are not in place. Aldrich, Nguyen and Maclaren, and Yang predict that strong civil society can enhance the political power of communities to sabotage unwanted projects and even engage in community development regulation after projects are sited, even in Communist states. Ishizaka et al. show how low levels of trust in, and sincerity of, institutions exacerbate risk perceptions. Kunreuther predicts that compensation is likely to increase support for projects partially, but that it is likely to be treated as a bribe if compensation is offered after safety assurances. Shaw predicts that compensation is likely to increase support for projects where social and political pressure for opposition is not strong and where it does not crowd out civic duty. Quah and Toh show how changing discourses can weaken adverse moral principles and enhance the effectiveness of compensation.

All of the analyses in this book provide explicit (although sometimes implicit) prescriptions for more effective siting. Mitchell suggests that policy makers should be more sensitive to environmental justice issues, while Baxter highlights the benefits of a multicounty process in terms of achieving procedural and distributive equity and generating a greater potential for altruistic roles. Kunreuther and Yang point to the need to enhance governance of siting processes through transparency, high quality public input, and social participation. Lam et al. point to the importance of building trust and engaging in public dialogue in a more open and frequent way in order to manage distrust, while Chiou adds to that, suggesting that conflicts should be managed by legal institutions. Nguyen and Maclaren argue that decision makers should encourage greater levels of public involvement to minimise the costs of community opposition in the future after projects are sited. Ishizaka et al. argue that developers should encourage regular communication between citizens and local government and that they should adopt more transparent policies in regard to information and performance reports. Quah and Toh argue that policy makers should provide the public goods that are the preferred form of compensation in addition to risk-mitigation strategies. Shaw suggests the need to design compensation strategies to increase civic duty and reduce opposition through social and political pressure. Aldrich implores policy makers to take into account the characteristics of civil society and to employ softer (as opposed to coercive) strategies in dealing with intense opposition.

While implicit, the chapters highlight important criteria in terms of evaluating siting processes and outcomes. Mitchell and Baxter suggest the importance of distributive equity and environmental justice in evaluating the success of siting. Kunreuther, Chiou, Lam et al., Nguyen and Maclaren, and Ishizaka et al. point to the importance of trust in, and credibility gaps among, government and social institutions and local communities, public engagement strategies to manage risk and local government roles in managing communications with community interests as critical indicators in learning from siting experience. Aldrich suggests that evaluation of siting should consider explicitly the civil society and whether policy instruments used to manage conflict were creative and innovative enough to deal with siting conflicts in the context of the features of those societies. Quah and Toh, Shaw, and Kunreuther indicate that siting could also be evaluated by assessing the nature of compensation mechanisms (especially the use of public goods compensation), the extent to which a combination of compensation and mitigation worked to deal with conflict, and the relationships among compensation and morality, social pressure and altruism.

EXTENDING KNOWLEDGE FRONTIERS

The existing literature on siting has provided the intellectual foundation for developing broader recommendations on the effective siting of unwanted projects. As noted by Lesbirel, the dominant set of principles is embodied in the Facility Siting Credo. While there are other recommendations to approaching siting, they are essentially variants of the Credo. Sequential multistage siting incorporated the key elements of public involvement, negotiations, and processes to determine compensation requirements. Cooperative discourse approaches also incorporate key elements of the Credo, such as exploring different options using both public and expert input into siting processes. The only real difference between the Credo and stepwise siting is the emphasis in the latter on social learning. Much of the extant literature on siting has generated recommendations that are consistent with the Credo, such as seeking to develop more trust and the like.

Yet, there is a debate in the literature about whether the Credo is working effectively. Kunreuther et al. (1993) have tested the Credo empirically in the United States and Canada in 29 siting cases, both successful and unsuccessful. They provide evidence that the Credo is likely to be effective where communities volunteer to host facilities, trust between the developer and host community can be established, the community perceives the facility design to be appropriate and to satisfy its needs, and the public participates in the process and develops a view that the facility best meets community needs. Shaw (2005) argues that experiences in a wider range of nations suggest that the Credo has failed to deliver approved sites. He argues that a focus on policy instruments alone does not take into account whether decisions about employing those policy instruments can minimise the failure of political institutions, and proceeds to propose a federal system that is characterised by a number of functional, overlapping, and competing jurisdictions (FOCJ) and a decision process based on the Principle of Interest-Pay-Participation (PIPP). This debate about policy instruments and institutions needs to go further. It needs to test more rigorously the applicability of the Credo in a wider variety of social and political contexts and we are limited in our evaluation of the FOCJ/PIPP proposal until we can observe its effectiveness in real siting cases.

Whether we focus on policy instruments, institutions or both, the literature and the policy recommendations that have emerged from it have made two critical assumptions that conceptually have limited its effectiveness in terms of knowledge production and utilisation. The first assumption relates to the scope of the community. Much of the literature and policy recommendations assume that the expected spillover effects of large-scale developments are confined to the host community defined in simple jurisdictional terms. The Credo explicitly refers to the community in these terms. For instance, it notes the importance of seeking consensus through a voluntary process, achieving trust and fully addressing all negative aspects of the facility, ensuring safety standards will be met, all in the context of making the *host* community better off. Hence, the Credo defines the relevant community in terms of spillover effects on the *host* community.

The chapters in this volume suggest that spillover effects are not confined to jurisdictional boundaries (administrative boundaries of a specific city, town, or village) and, as a result, siting processes often involved nonjurisdictional stakeholders in neighbouring and other jurisdictions. Kunreuther highlights the importance of transboundary risks and argues for a siting authority that has jurisdiction over a broader geographic area. Baxter shows in a Canadian case that, given expected benefits, Ryley, a neighbouring community, opposed Swan Hill's getting a project. Furthermore, the developers avoided Ryley because it expected the siting process there to involve a multicounty plebiscite. Lam et al. highlight that, given prevailing climatic conditions, Hong Kong (SAR) does not site projects in communities in the eastern part of the region that residents in the western part of the region would likely oppose. Yang notes how, after conducting an EIA in a surrounding region, the provincial government decided to locate a chemical factory at an alternative location. Quah and Toh demonstrate that seeking to locate a casino in one particular locality in Singapore led to the emergence of a moral debate in the city-state as a whole. Shaw interestingly shows how plans to site a waste repository on an island in the Taiwan Straits ultimately led to resistance by Mainland China which forced the shelving of those plans.

The empirical observations suggest the need to redefine the scope of community in the siting literature. Defining community in jurisdictional terms may be conceptually inadequate in terms of gauging the nature, scope, and size of spillover effects that a project may impose and the conflict that may emerge. While host jurisdictions will always play importance roles in siting processes, other communities may also be crucial. They may include jurisdictions at different levels of government, for example, broader regional (state) and national jurisdictions. They may be defined by geography such as climatic factors that cause spillover effects to be felt more broadly than the host jurisdiction. Similarly, they may be defined by economics, such as processes involving the specialisation of production over a range of host jurisdictions. Politically, they may be defined in electoral terms where electoral boundaries at state and national levels will be often broader than simply the host jurisdiction. In some cases, the community may also involve interstate actors where expected social, political, economic, or strategic spillover effects cross into other nations. We need to pay more careful attention to defining community and assessing the scope of the relevant community in siting analyses and to incorporate that explicitly into our siting policy recommendations.

The second major assumption concerns the decision-making process and outcome. Much of the literature focuses on the conflict involved in reaching a decision over whether to site projects in a particular locality. The Credo points to voluntary and consensual decision processes and suggests that guaranteeing stringent safety standards, making the host community better off through compensation, and using contingency agreements will enable a decision to be made. It assumes that in making a siting decision all the relevant spillover effects have be managed and that all potential sources of conflict are eliminated. Hence, the Credo presumes that there is no real need to worry about post-decision spillover effects and conflicts in terms of siting processes and their outcomes.

Several analyses contained in this book clearly suggest that spillover effects can continue beyond the initial siting decision on whether to site projects. Baxter shows clearly that, despite quick agreement on the Swan Hill project, 50 percent of residents in Swan Hill and two neighbouring towns would oppose the project if it were voted on today, and Baxter notes significant intercommunity conflict. Lam et al. note that one significant issue in the Tuen Mun area in Hong Kong (SAR) relates to the continued stigma associated with having a concentration of refuse tips. Nguyen and Maclaren demonstrate how an inadequate landfill-management plan in Vietnam led to conflict and informal regulation where the community monitored firms, even in a Communist state. The literature needs to pay more attention to post-decision conflicts, including those during the construction, operation, and decommissioning stages and to build recommendations that inform policy makers how to best manage siting processes for the life of projects.

REFERENCES

Kunreuther, H., Fitzgerald, K., & Aarts, T. D. (1993). Siting noxious facilities: A test of the Facility Siting Credo. *Risk Analysis*, *13*(3), 301–318.

Shaw, D. (2005). Visions of the future for facility siting. In S. H. Lesbirel & D. Shaw (Eds.), *Managing conflict in facility siting: An international comparison*. Cheltenham, UK: Edward Elgar.